HOT AIR

HOT

JEFFREY SIMPSON,
MARK JACCARD,
and
NIC RIVERS

HOT AIR

MEETING CANADA'S CLIMATE CHANGE CHALLENGE

EMBLEM

McClelland & Stewart

From Jeffrey: To Tait, Danielle, Brook, and Wendy, sine qua non
From Mark: To Ingram, Kjartan, Torsten, and Sigbrit
From Nic: To Simone

Cloth edition published 2007
Emblem edition published 2008

Emblem is an imprint of McClelland & Stewart Ltd.
Emblem and colophon are registered trademarks of McClelland & Stewart Ltd.

Library and Archives Canada Cataloguing in Publication

Simpson, Jeffrey, 1949-
Hot air : meeting Canada's climate change challenge / Jeffrey Simpson, Mark Jaccard and Nic Rivers.

ISBN 978-0-7710-8097-5

1. Climatic changes – Canada. 2. Greenhouse gas mitigation – Canada. 3. Climatic changes – Government policy – Canada. 4. Greenhouse gas mitigation – Government policy – Canada. I. Jaccard, Mark Kenneth II. Rivers, Nic, 1976- III. Title.

TD885.5.G73S54 2008 363.738'740971 C2008-900936-3

We acknowledge the financial support of the Government of Canada through the Book Publishing Industry Development Program and that of the Government of Ontario through the Ontario Media Development Corporation's Ontario Book Initiative. We further acknowledge the support of the Canada Council for the Arts and the Ontario Arts Council for our publishing program.

Typeset in Electra by M&S, Toronto
Printed and bound in Canada

A Douglas Gibson Book

This book is printed on acid-free paper that is 100% recycled, ancient-forest friendly (40% post-consumer recycled).

McClelland & Stewart Ltd.
75 Sherbourne Street
Toronto, Ontario
M5A 2P9
www.mcclelland.com

1 2 3 4 5 12 11 10 09 08

CONTENTS

PART I

HOW WE GOT INTO THIS MESS

The Science and the Politics

CHAPTER ONE

Why a Warmer Canada Is Bad News

Canadians know they shiver in a huge, cold, northern country – except residents of Vancouver Island and British Columbia's Lower Mainland, who taunt fellow Canadians with tales of crocuses and daffodils in February. Anything, therefore, that suggests their country could get warmer, or is getting warmer, might strike Canadians as a blessing. What could be wrong with shorter, milder winters and long, hotter summers? If the climate is indeed changing, and more hot air arrives, then bring it on.

But because Canada is a huge, cold, northern country, a warmer climate does threaten to change it, especially its Far North. Because Canada is a huge country, climate change presents not just one threat that a smaller country might have to confront, but a variety of threats, most of them negative. Because the geography of a cold country like ours has been shaped by snow and ice, a diminution or dilution of snow and ice changes the vast Arctic, the glaciers and mountain snowpacks that feed rivers, lake levels, sources of fresh water, treelines, tundra. The climate, in other words, changes the geography. And because the biggest temperature changes will be felt in the northernmost and southernmost parts of the earth, a northern country such as Canada cannot escape the effects of global warming. All countries will be affected by climate change, with those already experiencing extremes of heat (India) and drought (Australia and parts of Africa) or those with low-lying territory (Pacific islands, Bangladesh) being particularly vulnerable. But a big, cold, northern country will feel the effects, too. Whether they realize it or not, Canadians are on the front lines of global warming.

Canadian scientists were among the first to underscore the perils of global warming to the planet (and Canada) more than two decades ago. But the general public missed the warnings. The evidence seemed inconclusive, and not much media attention was paid to the phenomenon. Politicians, when told of the problem, preferred discussion documents, round tables, and policies designed to minimize political risk.

The effects of global warming on Canada, however, have been apparent for some time, as a few examples will illustrate.

The Arctic Council's five-year study by 300 scientists, published in 2004, found a 4-degree increase in temperatures in the western Arctic from 1953 to 2003, and a 2- to 3-degree increase elsewhere in the Canadian Arctic. (All temperatures in this book are stated in degrees Celsius.) The results in the Far North have already been dramatic, and more drama is on the way: less snow and more wind, falling water levels in northern lakes and rivers, more insects, decreasing sea ice, shrinking tundra, thawing permafrost leading to more difficulty building and maintaining winter roads, ongoing threats to traditional aboriginal ways of life, and widening of the Northwest Passage.

Global warming is already affecting chunks of Canada's forests. For example, the mountain pine beetle has killed half of British Columbia's lodgepole pine and will probably have killed 80 per cent of it by 2013. The provincial government calls the infestation a "forest health epidemic." Only very cold winters – what we used to consider "normal winters" – kill the beetles that nest under the trees' bark. Warmer winters for a decade have produced a looming economic catastrophe for towns, mills, and people dependent on a supply of pine. The outbreak that started in a provincial park now covers 8 million hectares of land in central and southern British Columbia – an area more than twice the size of New Brunswick – and is spreading east across the Rockies into Alberta and north into the Yukon. David Coutts, Alberta's minister of sustainable development, said in October 2006, "I pray every night that Mother Nature gives us minus-30-degree weather. . . . We are at war with the mountain pine beetle. And this year, Alberta is the battleground."

A study by the Canadian Forest Service, released in September 2006, reported the pine beetle was spreading at an alarming rate, with outbreaks as far east as Saskatchewan. In July 2006, a strong wind blew millions of pine beetles into the area around Grande Prairie, Alberta, where they began attacking Jack pines and lodgepole pines. Foresters are deeply worried that the infestation could leap to the boreal forest that spreads across much of Canada.

Glaciers are receding faster than ever, with consequences for water levels in the rivers that emerge from them. Some glaciers in British Columbia's popular Garibaldi Park, for example, have receded by 75 per cent over the past two decades. Areas of the Canadian prairies and British Columbia that rely on meltwater are likely to experience more long-term water shortages as a result of less glacial melt. Agricultural soils that depend on rivers and creeks retaining certain water levels are likely to become drier and, with variations according to circumstances, less productive. But in some parts of Canada, a warmer climate and more carbon dioxide (CO_2) in the air might help agricultural production, although if the buildup is too great, agriculture could suffer.

Increased temperatures can contribute to more air pollution in urban areas, since smog remains closer to the ground at higher temperatures. Smog is already a significant problem in southern Ontario. The Ontario Medical Association has estimated that smog was responsible for as many as 5,800 premature deaths in 2005.

A 2003 study for the Canadian Council of Ministers of the Environment (CCME) found that from 1900 to 1998, average temperatures in southern Canada rose by 0.9 degrees. During the last 50 years of that period, the west and northwest parts of Canada got hotter. Only the northeastern corner of the country (eastern Baffin Island, northernmost Quebec, and Newfoundland and Labrador) became cooler. Generally speaking, the weather is warming less at the high end of the temperature scale than at the low end: fewer cold winter nights rather than more steaming hot summer days.

Higher temperatures evaporate more water from the earth's surface. This leads to more rain in rainy areas in rainy seasons, and more droughts in dry areas. Canadian precipitation levels have risen everywhere except in the southern prairies. In fact, precipitation has increased by between 5 and 35 per cent in most of the country since 1950, the CCME study found. Sea surface temperatures on the West Coast rose by between 0.9 and 1.8 degrees over the last century. Data from the East Coast is spotty, but tests off Cape Spear, Newfoundland, show no change, consistent with the finding that only in the most northeastern parts of Canada has the climate not warmed.

A warmer climate increases the unpredictability of weather patterns, bringing on more unusual conditions of the kind that Canadians have been experiencing in recent years. In 2006, as Environment Canada laconically reported, Canadians had plenty to "weather."

British Columbia opened 2006 with a record number of wet days in January. In November, another record-setting wet month for Vancouver, coastal rain (and huge snowfalls in the mountains) hit the province, causing extensive flooding, landslides, and a provincial boil-water advisory for millions of people in the Lower Mainland. Barely recovered from the November storms, residents faced another lashing in December with three storms that damaged dozens of homes, left a quarter of a million people without power, and toppled hundreds of beloved trees in Stanley Park. Sandwiched between these stormy periods was a much drier summer than usual, producing 50 per cent more wildfires.

Overall, 2006 was the second-warmest year on record across Canada, and the tenth consecutive year of above-average temperatures. The winters of 2005 and 2006 were the two warmest on record, 3.9 and 3 degrees warmer than normal. Winnipeg recorded its driest June and July and its warmest January on record. Edmonton had its highest temperature in 70 years. A record number of hailstorms cost millions in property and crop losses on the prairies.

Farther east, Toronto's Pearson International Airport experienced its warmest night ever. Powerful thunderstorms in Ontario and Quebec killed four people and left thousands of customers without power. Southern Ontario and southern Quebec had the wettest fall on record and one of the wettest years ever. Montreal endured its rainiest year.

One year does not make a pattern. But when that year is added to a longer series that shows the same broad trends, it provides further evidence that Canadian weather patterns are shifting, because the climate is changing.

HEADS YOU GET WARMER, TAILS YOU GET HOTTER

Dave Phillips, Environment Canada's ebullient senior climatologist, once said, "If we could predict the weather perfectly, it would take all the fun out of being Canadian." That quip might be a slight exaggeration, but Canadians certainly do talk a lot about the weather. Until recently, they talked less about the climate. But that is changing.

Meteorologists distinguish between climate and weather. Climate is the average weather in a region during a long period, say, 30 years. The climate of a region (and its weather in all of its different guises) includes not only temperature but also snow, rain, wind, fog, atmospheric pressure, sunlight, and a host of other variables.

Canadian meteorologists have been measuring these variables at hundreds of locations for over 60 years, and for over 150 years in some locations. They have a good idea of how weather in a given location varies throughout the year. Analyzing these measurements over long periods of time gives them a picture of the climate of Canada.

Just because a wealth of data has been collected over the past 60 years does not mean weather can be predicted with precision over the next 60 years, or even over the next week. Instead, what we can do is predict the likely weather in the future – provided the underlying climate does not change. So we intuitively know that the temperature is not likely to fall below freezing on a summer day in Windsor, or that the temperature would not rise above freezing during a winter night in Yellowknife.

Anomalies, however, happen all the time because the weather is such a complicated and chaotic system. For example, the average daily high temperature in Toronto in June is 23.5 degrees, and the average daily low is 14.8 degrees. Variations occur around the averages. A daily high of 28 degrees or one of only 15 degrees would provoke little comment. But on June 30, 1967, Toronto's daytime high temperature reached 36.7 degrees, while in the same month 122 years earlier, the temperature had dropped to –2.2 degrees.

Those temperatures would certainly stir comment. But anomalies such as those two days, surprising as they were, could have been expected, because Toronto's weather, like weather elsewhere, is influenced by random processes.

Understanding random processes is helped by thinking of the familiar act of tossing a coin. If a coin gets tossed 10 times, the expected outcome would obviously be five heads and five tails. But we would think nothing of it if the outcome were six heads and four tails, or seven heads and three tails. If the coin gets tossed enough, we would expect the extreme situation of 10 heads in a row only once every 1,000 times.

Imagine, however, that in repeated tosses, all the sets of 10 came up with more than five heads. This result would not be surprising at first, because tossing the coin is a random process. But if more than five heads happened all the time, we would probably start thinking that something was amiss with the coin or with the way it was being tossed. In tossing a set of 10 coins 30 times, for example, it would be surprising if more heads than tails appeared 30 times. In fact, this outcome would happen randomly just once in every 5 trillion times. Even flipping the coin fairly quickly, a person would have to spend about 5 million years continuously flipping to witness this outcome.

This outcome is what scientists have started to observe with the earth's weather. Some winters are going to be warmer than normal, and some will be cooler. That's the normal variation in the weather. But we did not expect, until now, long periods where the weather is warmer than normal. And we would never expect it unless the underlying system – the earth's climate – has changed. Something quite fundamental was therefore occurring when the year 2006 featured warmer-than-average temperatures across the globe for the 28th consecutive year. All 10 of the warmest years on record have occurred since 1990, and the three warmest years on record have all occurred since 1998. The probability that this warming is a natural outcome of variable climate gets closer to zero with each additional year of warmer temperatures. The warming is not a random event – think of the coin toss analogy – but a sign that the climate has fundamentally changed, and will keep on changing.

HUMAN IMPACT ON CLIMATE

So if Canada's and the world's temperatures are rising, the question is why. Who and what are responsible, in Canada and internationally? The answer, although much in the news recently, is actually an old one: human activity.

In the late 1800s, the Swedish chemist Svante Arrhenius, building on earlier scientific inquiries, first suggested that humans might play a part in a changing climate. His work dealt with the sensitivity of the Earth's temperature to the concentrations of carbon dioxide in the atmosphere. Working with a model that climatologists today would consider ridiculously simple and naïve, Arrhenius estimated that the Earth would warm by 5 to 6 degrees if atmospheric CO_2 was doubled, not far off today's more informed predictions of around 3 degrees.

With advanced computer models and hundreds of scientists studying the climate, two things now appear highly likely: first, that the sharp increase in emissions of greenhouse gases (carbon dioxide, methane, and others) has resulted from human activities; and, second, that these emissions have warmed, and are warming, the planet. The most comprehensive and conclusive scientific assessment of the evidence has come from the Intergovernmental Panel on Climate Change (IPCC), an international body of scientists charged with summarizing evidence from the scientific community and reporting to the United Nations. The latest IPCC report on the science of climate change, released in February 2007, debunked the skeptics' assertion that climate change came from some source other than human activities, which are referred to as "external forcing." Said the IPCC, "It is extremely unlikely that global climate change of the past fifty years can be explained without external forcing, and very likely that it is not due to natural causes alone." Average northern hemisphere temperatures, the IPCC reported, during the second half of the 20th century were "very likely higher than during any other 50-year period in the last 500 years and likely the highest in at least the past 1,300 years." The recent Canadian warming pattern, not surprisingly for a northern hemisphere country, mirrored the IPCC finding that "eleven of the last twelve years (1995–2006) rank among the twelve warmest years in the instrumental record of global surface temperatures."

Climate sensitivity is indicated by the increase in surface temperatures resulting from a doubling of carbon dioxide in the atmosphere. Before the industrial age, concentrations of CO_2 were about 280 parts per million in the atmosphere; today they are more than 380 parts per million, and the rate of yearly increase is growing. Methane concentrations have increased about 150 per cent since the pre-industrial age. The primary reason for the higher concentrations: fossil fuel use. A secondary reason: changes in land use. With what the IPCC scientists called "higher confidence" than in previous reports, they forecast, among other results, that warming would be greatest over land and at northern latitudes (places such as northern Canada), and they predicted less snow cover, more permafrost thawing, shrinking sea ice, increased frequency of heat waves and heavy precipitation, and more violent storms.

Sir Nicholas Stern, a respected former chief economist and senior vice-president of the World Bank, released in 2006 a report for the British government summarizing the state of climate change science and analyzing the economic consequences of global warming. His review of the science followed the IPCC's findings. What made headlines, however, was his economic conclusion: that the cost of abating climate change would be 1 to 3 per cent of global gross domestic product, whereas doing nothing about climate change would produce a catastrophic decline of 20 per cent in global GDP. "Our actions now and over the coming decades could create risks . . . on a scale similar to those associated with the great wars and economic depression of the first half of the twentieth century."

Stern's warnings ricocheted around the world. They also produced swift and serious criticisms. William Nordhaus of Yale University, possibly the most influential climate change economist of recent decades, dismissed Stern's conclusion, saying that Stern had wrongly accounted for future costs. A dollar in the future is worth less than its value today. Most economists, including Nordhaus, would assume a "social discount rate" of 1 to 3 per cent over the long term, whereas Stern assumes a rate of 0.1 per cent, effectively zero. Stern's assumption leads him to propose massive one-time investments as soon as possible for what would be small improvements in a hundred years. Nordhaus and others propose more sensible solutions that would see climate change policies start slowly today but increase in stringency over many decades. That is the kind of long-term approach – start effective policies now and in due course make them more stringent – that a big, cold, northern country such as Canada should pursue.

THE HEAT EXCHANGE BALANCING ACT

The earth's climate depends on the delicate balance between heat leaving and heat entering the atmosphere. Essentially, all of the outside energy that reaches earth comes from the sun. The amount of that energy is immense. On average, for every single square metre of the earth's surface, the sun delivers 342 watts of energy – equivalent to the amount consumed by about six standard household light bulbs. About a third of that, or two light bulbs, is immediately reflected back to space by the earth's atmosphere, clouds, and the earth's surface. The remaining 235 watts per square metre, or four light bulbs, heats the earth's atmosphere and surface before also returning to space.

Heat naturally wants to flow upward from the warm earth to space, where temperatures can be –270 degrees, in the same way that heat flows from a hot cup of coffee into the cooler air of the room around it. The atmosphere acts as a kind of insulator, keeping the earth's temperature more or less stable at a life-sustaining average 14 degrees.

The atmosphere contains about 13 trillion tonnes of water vapour – enough to fill Lake Superior if condensed. Carbon dioxide is also present in the atmosphere in a tiny quantity – for every 50 molecules of water vapour there is one molecule of CO_2 – but carbon dioxide is a wonderfully efficient insulator. One molecule of CO_2 insulates the earth much better than a molecule of water vapour.

Water vapour, carbon dioxide, methane, and other gases came to be called greenhouse gases (GHGs) because of their ability to trap heat. It stands to reason, therefore, that if there is more carbon dioxide (the most common offender) and other insulating gases around to warm up the atmosphere by slowing the outward flow of heat, the earth's surface temperature is going to increase. And that is what has been happening, just as a garden greenhouse lets the sun's rays in but keeps the resulting heat from getting out.

Changes in CO_2 levels and earth temperatures have been strongly correlated over hundreds of thousands of years. Using data collected from Greenland and Antarctica, scientists have demonstrated a nearly perfect correlation between levels of CO_2 in the atmosphere and global temperatures over the last 400,000 years. Levels of CO_2 have fluctuated between 180 and 280 parts per million in the atmosphere. When levels were around 180 parts per million, the earth's temperature was 5 degrees cooler than at present. Glaciers in those days extended from the poles to cover large parts of the earth. Carbon dioxide levels of 280 parts per million corresponded to periods when temperatures increased, and glaciers retreated toward the poles. Remember, therefore, when thinking about climate change that just 5 degrees represents the difference between average temperatures today and the Pleistocene Ice Age, when glaciers covered much of North America.

What do these changes have to do with human activity? Over the last two centuries, and especially in the last 50 years, humans have burned enormous quantities of coal, oil, and natural gas for all sorts of purposes. Reputable estimates suggest that over the last two centuries, humans have released into the atmosphere more than 1 trillion tonnes of carbon dioxide by burning fossil fuels. This amount has about the same mass as the water in roughly 500 million Olympic-sized swimming pools, or one swimming pool's worth (by mass) of CO_2 for every dozen people alive on earth today.

When these fuels are burned, the products of combustion are atmospheric insulators, namely water vapour and CO_2. Compounding the problem, carbon dioxide sticks around for a long time after being released into the atmosphere – an average of 100 years – before being absorbed by plants or oceans. Added trouble comes from human activity such as deforestation and urbanization, which have decreased the amount of carbon naturally absorbed by the earth's surface. And CO_2 is not the only GHG to worry about. Methane is 21 times more powerful than CO_2 as an insulator. Human activities, including our methane-belching livestock, are releasing ever more methane into the atmosphere. We are also causing increased emissions of nitrous oxide and halocarbons, extremely powerful greenhouse gases that can reside in the atmosphere indefinitely.

CO_2 emissions have been accelerating over the past half-century, and, to make matters worse, the other, more powerful GHGs have been rising even faster. Emissions are now forecast to rise even more rapidly in the next 50 years, unless remedial action is taken. Emissions from developed countries are projected to increase somewhat over the coming decades, while those from developing countries, led by China and India, are projected to skyrocket. From today's 26 billion tonnes of CO_2 equivalent (the theoretical sum of all GHGs converted, for comparison, into an equivalent amount of CO_2), global emissions are projected to be about 40 billion tonnes by 2030, and to grow even faster thereafter. Researchers at the Massachusetts Institute of Technology forecast that emissions will increase more than threefold to over 80 billion tonnes by 2050, a figure consistent with predictions in other studies. With these emissions will come, of course, even higher concentrations of GHGs in the atmosphere, with more global warming. How much warming? We already know that the 1990s were the hottest decade in 140 years of global climate records. International models suggest a climate that could be 2 to 6 degrees warmer over the next century. Remember that a decrease of just 5 degrees plunged us into an ice age.

WE'RE SUCH A TINY PART OF THE WORLD – WE'RE NOT RESPONSIBLE

Critics of serious domestic action against global warming argue that since Canada produces only 2 per cent of global emissions, we are not really a major cause of the problem. Yes, they admit, we should probably do something, in due course and at our own speed, but any Canadian action will be of only marginal overall benefit, given how little Canadian emissions contribute to overall climate change. This line of argument has fortified those who, while grudgingly coming around to conceding that human activity has produced climate change, have sought to slow down or frustrate Canadian action. It's been a favourite argument of many business leaders and scattered media commentators. That Canada has about 0.05 per cent of the world's population of 6.5 billion people but emits 2 per cent of all GHG emissions, making Canada one of the world's largest per capita emitters, does not strike the critics as important.

What does this 2 per cent represent? In 2005, Canada's GHG emissions were 747 million tonnes, a 25 per cent increase over the 1990 level of 596 million tonnes, whereas our international commitment called for a reduction of 6 per cent from 1990 levels by 2008–12, to 566 million tonnes. The so-called "Kyoto gap" – between what we promised to do at the Kyoto conference and our actual emissions – in 2005 stood at 184 million tonnes, and is forecast to reach 270 million tonnes by 2010.

Only Turkey, Spain, and Portugal had faster rates of emissions increases among member countries of the Organisation for Economic Co-operation and Development (OECD). Even the United States did better than Canada. For the most emissions per capita, Canada ranked third out of 29 countries in the OECD. In absolute terms, Canada's total carbon dioxide emissions were eclipsed in the OECD only by the United States, Japan, Germany, and Britain. Canada's record, by any standards, has been among the worst.

The "only 2 per cent of the world" argument, if followed logically, would freeze Canadian action across the entire range of international initiatives. If Canada had followed that logic at the beginning of World War II, we would have never joined the fight against Germany and Italy, since Canada had only a small population with limited military means. Canada's foreign aid, in Africa and elsewhere, cannot reverse conditions of poverty, malnutrition, HIV/AIDS, and other woes, yet Canadians do not argue that because their assistance alone cannot solve these problems, the country should do nothing. Canada remains a member of the North Atlantic Treaty Organization (NATO), with 2,500 of our troops in Afghanistan, but whatever arguments might lead Canada to consider withdrawing them, the weakest one is that we should do so because the country cannot win the fight alone, since nobody assumes that we can and nobody is asking that of us.

Canada is among the world's great joiners, despite its modest population size and relative geographic isolation from other continents. Canada helped to found the United Nations, NATO, the International Monetary Fund, the World Bank, and a host of other international organizations, including the Commonwealth, and la Francophonie. Canada never took the position that because it was a small country compared with many others, it should therefore not contribute ideas, money, and commitment to resolve world problems and enhance international stability – especially since a realistic understanding of the world illustrated that the amelioration of problems and the enhancement of stability were in Canada's own long-term self-interest. If the argument had been advanced that Canada's contribution alone was too small to affect outcomes, then Canada would have rolled itself up behind the oceans and lived a rather more carefree, but irresponsible, national life.

Canada is an advanced industrial country with an enviable standard of living based in part on development and use of energy resources. As such, Canada is part of the group of developed countries that have for a long time been emitting GHGs, thereby contributing to global warming. True, some developing countries are now contributing to the buildup, but the obligation of developed countries to make changes in their emissions patterns can be understood by considering this figure.

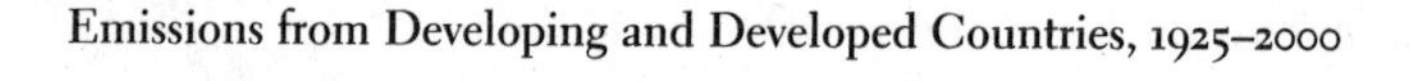

Emissions from Developing and Developed Countries, 1925–2000

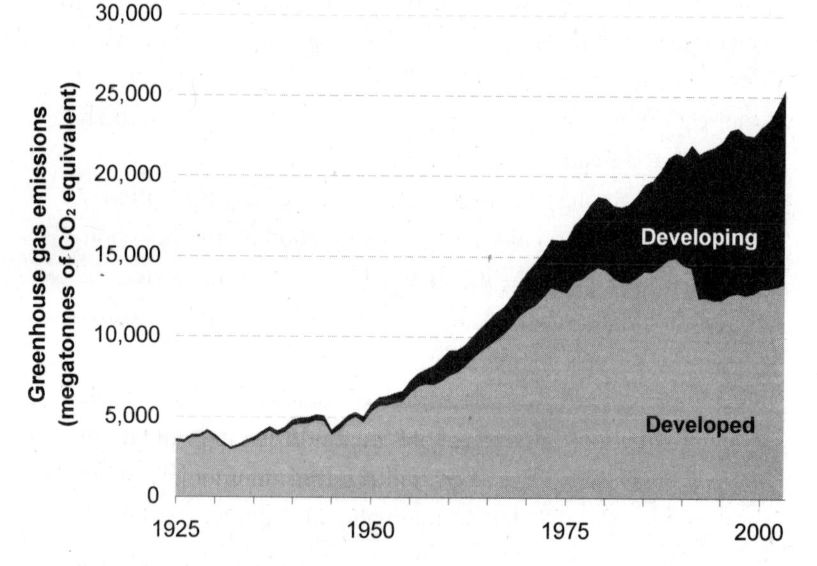

Source: Data from Marland, G., T.A. Boden, and R.J. Andres. 2006. "Global, Regional, and National Fossil Fuel CO_2 Emissions." In *Trends: A Compendium of Data on Global Change*. Carbon Dioxide Information Analysis Center, Oak Ridge National Laboratory, U.S. Department of Energy, Oak Ridge, Tenn. Only includes CO_2 emissions from fossil fuel combustion, cement production and gas flaring.

THE TRAGEDY OF THE COMMONS

The 2 per cent argument illustrates the classic tragedy of the commons. Aristotle wrote about the phenomenon, "That which is common to the greatest number has the least care bestowed upon it. Every one thinks chiefly of its own, hardly at all of the common interest, and only when he is himself concerned as an individual." The atmosphere is a classic "commons," like oceans, owned by no one but available to all. Free access and unrestricted demand can ultimately threaten a commons with degradation. Since no one user has done enough identifiable damage, no incentive exists either to punish the user or to prevent the user from continuing to behave in the same damaging way. It is therefore rational behaviour for each user to carry on regardless – until it becomes too late and the resource begins to run out.

Canadians know all too well about the tragedy of the commons. The East Coast cod fishery was fished to near extinction, with devastating economic effects, especially in Newfoundland and Labrador. Assigning blame – to inshore or offshore fishermen, or to foreign fleets – is a matter for passionate, ongoing debate. But the evidence is strong that the stocks were overfished, despite warning signs. The fishery was unable to regulate itself properly and avaricious foreigners could not be restrained. All the fishing interests preyed upon the common resource. No one fisherman, company, or foreign country entirely caused the collapse, and therefore no single interest saw fit to act before the collapse. Indeed, as the initial signs of the stocks' decline emerged, various actors dependent on the resource finger-pointed their way to inaction. Inshore fishermen blamed offshore fleets. Offshore fleet owners blamed faulty science, insisting their captains were finding fish and suggesting the small inshore operators just wanted more for themselves. Everyone in Canada could agree that greedy foreigners – Spaniards and Portuguese, mostly – were the real culprits, even though the vast majority of the stocks swam inside Canada's 200-mile zone and were off limits to foreigners in that zone. Canadians caught most of the fish, but it was expedient to blame foreigners who were doing their bit, to be sure, to wipe out the stocks. The provincial government blamed the federal government for poor monitoring; the federal government blamed the province for having licensed too many fish plants, thereby keeping too many fishermen on the water against what the stocks could bear. Denial and delay, followed by more denial and delay, prevented the timely, firm action required to save the stocks.

TALKING CHINA AND OTHERS AROUND

Climate change deniers and skeptics insist that Canada should take no action, or very little action, to reduce its greenhouse gas emissions until countries such as China and India reduce theirs. This position sounds logical. In 2001, U.S. president George W. Bush used it, among others, to justify why he refused to seek ratification of the Kyoto Protocol, and Australia under Prime Minister John Howard followed in Bush's footsteps. The absence of China and India from Kyoto made that position more compelling, until you look at the issue from their perspective.

GHG reduction requires those countries and industries that move first to accept higher initial costs and to introduce slightly more risky technologies. The poor people of the planet have limited access to the energy services enjoyed in richer countries. Average per capita electricity consumption in Bangladesh is just over 100 kilowatt hours per year and in China, 1,000 kilowatt hours, compared with over 10,000 kilowatt hours in Canada. Globally, almost 2 billion people are without electricity and another 2 billion get only sporadic service.

The Chinese are understandably focused on expanding their electricity system with what resources are available. China has huge coal reserves but little natural gas. The cheapest and least risky option for China, therefore, is to burn coal in conventional electricity generating plants – the very ones that are contributing to a significant share of human-produced GHG emissions. Every week, the Chinese open a new coal-fired plant, a serious menace to the global climate. For the last five years, China has been installing almost 50,000 megawatts per year of new conventional coal-fired generating plants – the equivalent of twice the entire annual generating capacity of Quebec – and the estimate for 2006 is an astounding 100,000 megawatts.

It all comes down to a basic fact of life: it's hard to persuade another person, let alone another country, to alter course if you remain on the same course you advise others to abandon. Lectures to the Chinese or the Indians, or residents or governments of any other country, about reducing GHG emissions lack credibility if the country doing the lecturing has not mended its own ways. If Canadians do little or nothing to reduce our own GHG emissions, whatever moral authority or political credibility we might have evaporates, and with it goes any chance of influencing China, India, and other developing countries.

Those countries will likely respond best not so much to moral suasion but to carrots and sticks. On the carrot side, we need to demonstrate what technological and resource options are available, and to manufacture these at the lowest possible price with the fewest mechanical risks. This requires us to accept the technological risk first, but as with all business, risk properly taken can produce opportunities to sell and make later profits. Canada might nudge developing countries along by assisting them to purchase this technology, although behind this carrot will likely lie a stick.

Why did the Chinese respond as they did to the Kyoto Protocol in 1997? Prior to Kyoto, the Chinese had insisted, along with many other developing countries, that climate change was the fault of the wealthy industrialized countries. They should solve the problem they were creating, and at their own expense. China therefore refused to commit to any GHG reductions.

Once Kyoto was signed, however, China acted differently, while still insisting it was under no international obligation to do anything. China removed subsidies for the coal industry, established mandates for renewable electricity generation, created a state-owned company to finance research into carbon capture and storage for coal-bed methane production, tightened energy efficiency standards, and tightened vehicle efficiency requirements.

Chinese officials will admit privately that they saw the writing on the wall, at home and abroad. Environmental degradation was becoming a serious domestic problem, leading to protests in various places across China that the Communist Party found unsettling. China also feared that industrialized countries moving against global warming would not sit idly by while China's emissions wiped out gains elsewhere. China would be widely criticized if it did nothing. For a country so dependent on exports, international opprobrium was not what the Chinese wanted or needed. The stick was the possibility, even the probability, that countries would in due course begin to use economic sanctions against global warming laggards such as China. But, of course, if the Chinese observe countries such as Canada not doing much, then the stick argument (and the carrot one) becomes very weak indeed.

As for the United States, that country is also taking action, not within the Kyoto Protocol to be sure, as long as George W. Bush remains president, but at the state, municipal, and congressional levels, to begin reversing the upward trend in emissions. A growing number of business leaders, governors, and mayors are acting – not jawboning but acting – to implement policies to drive down emissions, with California leading the way. American business, outside the fossil fuel and coal industries, sees burgeoning economic opportunities – a.k.a. profits – in new technologies and energy savings, domestically and internationally.

In Australia, another Kyoto refusenik country, torrid summers, extensive drought, and voracious wildfires have switched public opinion in an already hot country from shrugging about global warming to almost universal demands for more action than Prime Minister John Howard had contemplated. Howard, a long-time climate change skeptic, was forced by public opinion to begin taking climate change more seriously after five years of drought produced what Australians call the Big Dry, although most of his measures were adaptations to the drought through more careful water management rather than stiff measures to reduce emissions and therefore mitigate climate change. Emissions actually rose in Australia by 25 per cent over 1990 levels by 2004, although by claiming credit for reforestation and for agricultural and forest land that absorbed carbon, the Howard government said emissions had risen by only 5 per cent, on target with the country's commitment to slow the increase of GHG emissions to 8 per cent over 1990 levels.

WHAT ARE OUR PROBLEM AREAS?

Where are the increasing emissions coming from in Canada? The Canadian government had to answer that question, among many others, in considerable detail for the United Nations in order to comply with its international obligations. The latest figures are from 2004, but they show trends that continue to this day. Electricity and heat generation, fossil fuel production, mining, transportation, farm animals, and waste all released more emissions between 1990 and 2004; manufacturing, construction (except mining), and chemical and metal producers produced fewer.

Alberta has the highest per capita emissions, Quebec the lowest. Overall emissions grew fastest in Alberta and Saskatchewan, largely owing to expanding oil and gas production. These provinces have by far the highest per capita emissions. Alberta's absolute emissions are comparable to those of Ontario, as the next figure shows. Alberta, with about 12 per cent of Canada's population, accounts for 31 per cent of Canadian emissions. This means that although GHG reduction must occur right across the country, the reductions in these two western provinces must be greater if emissions are eventually to fall below 1990 levels.

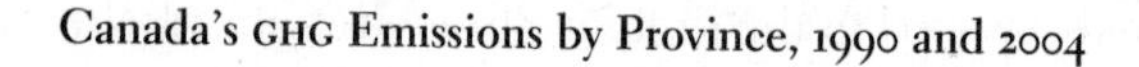

Canada's GHG Emissions by Province, 1990 and 2004

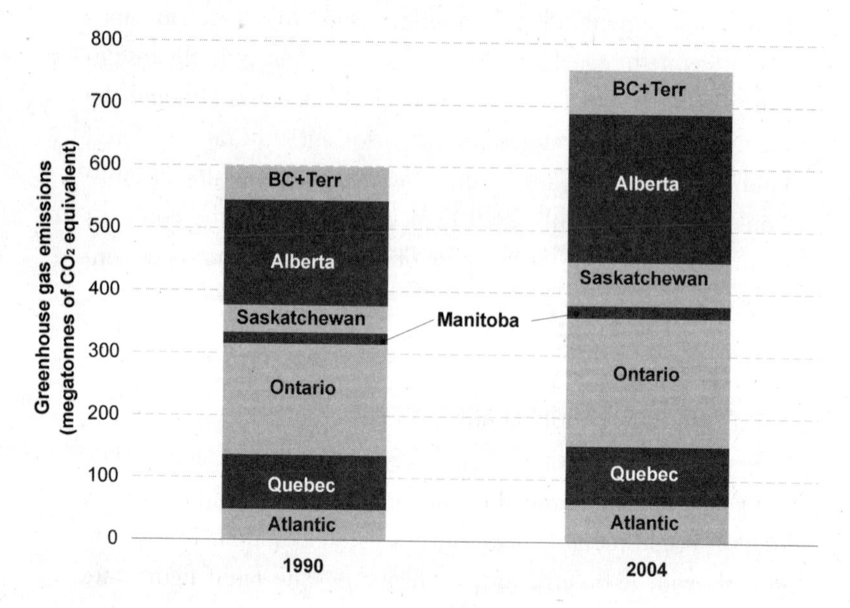

Source: Adapted from Environment Canada GHG Inventory, 2006.

Regional considerations often lie at the heart of national decisions in Canada. So it is with GHG reductions. Nor surprisingly, views about how aggressively to reduce emissions, how it can be done, and who should pay vary widely and depend upon the region. Judging by the governments they keep electing, most Albertans don't want to close down their fossil fuel industry, place a moratorium on oil sands development – as suggested by former premier Peter Lougheed – or even slow down the pace of growth. Some Canadians elsewhere argue, however, that Alberta (and Saskatchewan) must bear the brunt of the cost of cutting emissions since they produce by far the largest amount of carbon per capita. Others argue that all Canadians must share the economic burden of reducing GHGs.

Any policy that targets one or two provinces will fail, and rightly so. Targeting only Alberta and Saskatchewan would make it impossible to secure their agreement to any national program, and that agreement would be essential, because in Canada provincial governments are constitutionally responsible for natural resources. All Canadians, wherever they live, are responsible for GHG emissions because these are driven by the energy they consume in homes, vehicles, and commercial outlets – and by the energy required to produce and deliver the diversity of products they consume. Lower GHGs result from the production of energy than from its consumption.

Nor would any policy be effective if it targeted just one or two sectors of the Canadian economy, as the following figure shows. Seventy-three per cent of Canada's total emissions come from burning fossil fuels, mostly for electricity, heat, and transportation. Electricity generation (power and heat) is a large GHG emitter.

"Upstream" emissions from oil and natural gas production are also important. A major source of these emissions is energy use and methane leaks in the transport and processing of raw natural gas into market-ready natural gas. Overall, these upstream emissions have grown by a dramatic 48 per cent in the last 15 years, driven to a significant extent by expanding oil sands production, greater domestic and American demand for transportation fuel, and an increasing appetite for natural gas in domestic and American markets.

In the transportation sector, overall emissions grew by 35 per cent from 1990 to 2004, despite a reduction in emissions from cars. How could that be? Because while emissions from cars declined, consumers bought more trucks and SUVs, and manufacturers reclassified vehicles from cars to trucks because these faced less stringent tailpipe emissions standards. The result produced a doubling of emissions from SUVs and light trucks – which is not surprising, since on average these vehicles emit 45 per cent more GHGs than cars. Heavy-duty diesel trucks were the other vehicle category where emissions almost doubled.

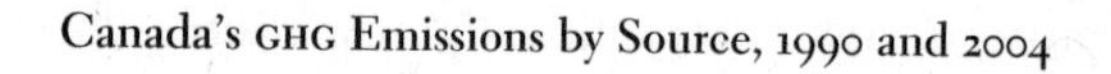

Canada's GHG Emissions by Source, 1990 and 2004

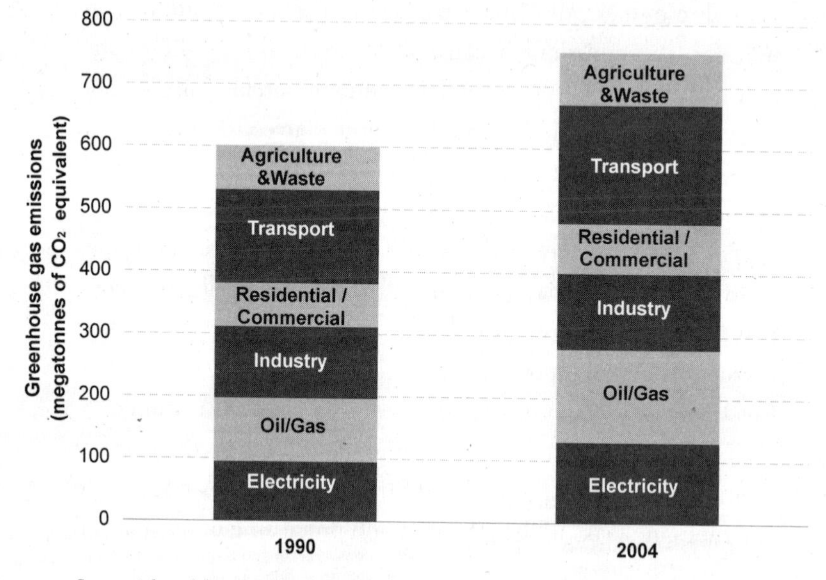

Source: Adapted from Environment Canada GHG Inventory, 2006.

IN CANADA, THE CANARY IS COUGHING

A cynic, or a practitioner of realpolitik, might find it easier to adopt the argument that Canadian action is not required if nothing bad were happening in Canada. It would be comforting, if hardly defensible for a responsible international country, to do nothing if problems appeared elsewhere but not at home. Under these circumstances, a country might adopt an "I'm all right, Jack" approach, and pretend the challenge of climate change bothered other countries but not ours.

This big, cold, northern country, however, has not escaped, cannot escape, and will not escape the consequences of climate change. These consequences are already apparent and will likely grow more pronounced in coming decades.

For example, an area of sea ice slightly larger than New Brunswick has disappeared in the western Arctic in the last three decades. More recently, in just a couple of hours in the early afternoon of August 13, 2005, a giant ice shelf about two-thirds the size of the city of Vancouver broke clear from the coast of Ellesmere Island. Dr. Warwick Vincent of Laval University was incredulous: "It was extraordinary. This is a piece of Canadian geography that no longer exists." Scientists think this event was the largest breakup in 30 years of continuous monitoring of Canada's Arctic.

Scientists estimate that Arctic sea ice is retreating at a rate of about 3 per cent per decade. An American study suggested the remaining ice had become 45 per cent thinner, while a more recent Canadian Ice Service study put the reduction at about a quarter of that estimate. The two agreed the ice was thinning, which means more will disappear. The disagreement was about how much and how fast.

Changes in sea ice may make it harder for species such as polar bears and seals to survive. It makes hunting more challenging for

Inuit. But there is something yet more ominous: less sea ice can lead to more climate change. Ice usefully reflects much of the sun's energy. With less ice, more of the sun's energy is absorbed into the water, so that temperatures rise faster.

With less sea ice, the fabled Northwest Passage could open up. It is far too early to predict when, or even if, this passage could be available to commercial shipping. But if it were opened, additional economic opportunities would have to be balanced with extra environmental risks from spills in the fragile Arctic. The possibility of navigation has already prompted concerns about asserting Canadian sovereignty – an assertion contested by countries with bluewater navies such as the United States and Britain – with promises of more research, manning of field stations, and building icebreakers.

All these Arctic trends foreshadow problems throughout the lower latitudes. As the University of Victoria Arctic researcher Terry Prose has observed, "We often refer to the Arctic as the bellwether, or the canary in the coal mine. Well, that canary is coughing."

Lakes are generally freezing later and thawing earlier. Ontario's Lake Simcoe is freezing 13 days later than 140 years ago and thawing 4 days earlier. For Swift Current Creek in Saskatchewan, freeze-up comes 24 days later and breakup 14 days earlier. This pattern of later freeze and earlier thaw is consistent with an international study of 39 rivers and lakes in Europe, Asia, and North America. One problem caused by this new pattern is the difficulty of building and maintaining ice roads that are vital for many northern communities.

Many lake and river temperatures are slowly rising. Fish species get introduced into new bodies of water because of changed water temperatures that can change parts of the ecosystem. For fish that spend at least a portion of their lives in fresh water, higher river and lake temperatures can lead to increased mortality and reduced growth. This could have important implications for salmon on both of Canada's coasts.

Glaciers around the world are shrinking. Since the greatest warming in Canada has been in the northwest, it's not surprising that the 1,300 glaciers on the eastern slopes of the Rockies are now about 25 to 75 per cent smaller than in 1850. Prairie farmers who depend on the water flows from rivers such as the South and North Saskatchewan, which rise in those glaciers, have every reason for concern. Again, the Canadian pattern mirrors the worldwide one. A study for the World Resources Council found the total size of the world's glaciers had shrunk by 12 per cent during the 20th century.

Warmer weather is changing certain Canadian ecosystems. New species are being seen in various parts of the country, sometimes in places where they acclimatize nicely, sometimes in areas where they confront conditions that make their life more precarious. Meanwhile, as every city dweller understands, more GHGs trap more heat, and that in turn leads to more smog conditions in Canadian cities.

Coastal areas, especially in Atlantic Canada, are more at risk from higher sea levels associated with climate change. Scientists have found that ocean levels have risen by 10 to 20 centimetres in the last century. Estimates from the Intergovernmental Panel on Climate Change suggest that sea levels will likely rise by between 10 and 90 centimetres over the 21st century. As a result, low-lying cities such as Charlottetown and other towns along the Prince Edward Island coast will be more exposed to sea surges. It is even possible that, without substantial reinforcement, dikes in the Bay of Fundy could be breached within the next 50 years.

These Canadian coastal risks pale beside those confronting low-lying islands and states with coastal areas just above sea level. They would see vast areas disappear with just a 1-metre sea level rise. Some estimates suggest that as many as 200 million people could be displaced by sea level rise over the coming century. The rise will come from glacial melt and the fact that water volume expands at higher temperatures, but these changes occur very gradually so that sea levels won't fully respond to a warmed climate for many centuries. But over the very long term, substantial melting of the Greenland ice sheet alone could raise sea levels by up to 7 metres.

Some Canadian agricultural areas will benefit from a longer growing season, less frost, and more abundant rainfall; others will experience more droughts and insects and lower yields. The mountain pine beetle is one prominent and devastating example.

These developments, among many, illustrate that no part of Canada – this big, cold, northern country – has been left untouched by climate change. Nor will any remain untouched in the decades to come. If anything, these effects will become more pronounced.

Various business groups and climate change skeptics or delayers of action have warned of dire economic consequences that will strike the Canadian economy if serious measures are taken against climate change. They have not – nor has any group, including governments – tried pulling together a comprehensive estimate of the costs to Canada of lack of action.

Beyond the skeptics and the delayers are millions of citizens for whom climate change has been an abstract problem divorced from their daily preoccupations. Part of the problem in attracting their attention concerns timing. Acting seriously against climate change would impose short- to medium-term costs on companies, governments, and individuals; not taking action will result in costs and risks over a much longer term. The immediate costs tend to be traceable and measurable; the long-term gains tend to be diffuse, and the long-term costs, by definition, remote and uncertain.

Even if people feel that doing something about climate change is important, their attention is easily and understandably diverted to more immediate concerns about health care, taxes, jobs, personal security, education. Most people are likely to display only a passing interest in problems lacking a direct link between action today and results tomorrow. Traditionally, cynical politicians count on this.

There are also uncertainties about climate change – fewer and fewer with each new study, but nonetheless estimating risks is a challenge. Critics therefore have argued that we should not act until we are certain about the precise dimension of the risk. More study before action has been a constant refrain of climate change skeptics and delayers for two decades in Canada.

Curiously, that is not how we address other risks. We follow a rather standard procedure with other hazards in our daily lives and businesses. If many independent experts tell us that a risk is significant, we do not usually pretend that we know more than the experts and act as if the risk were zero. We put on seat belts in case our car crashes, or the police detect us without one – two kinds of risk. We spend money on fire alarms, fire extinguishers, fire insurance, heat-activated sprinklers. We take these precautions even though we are far from certain that our house will catch fire.

When people argue against taking serious action, they need to explain why we would treat climate risk any differently than other types of risk. We are definitely not certain exactly how we are affecting the climate. But scientists are virtually unanimous now in telling us that it is highly likely that we *are* changing the climate, and that the change will be negative. With such uncertainty, no one can guarantee that action will prevent disaster. But the experts tell us it should change the probability.

In other words, if we approach this problem like other problems of risk, uncertainty alone is not a reason for inaction, although there is plenty of room for debate about the appropriate timing, cost, and effectiveness of actions. Some approaches – such as those proposed in this book – are better than others for cost and effectiveness. Certainly those that have been tried thus far have failed.

Delay is always an option, as is the avoidance of difficult decisions. It is easy for companies worried about maximizing the returns to shareholders today in a competitive marketplace to close their eyes to problems down the road. It is easy for individuals to say that their lifestyle decisions will have only an infinitesimal effect on climate change, and so carry on as before. And it is easy for politicians, caught up in a short-term political cycle of elections and an even shorter-term vortex of media attention, to put off today and tomorrow actions whose positive effects might not be apparent for decades.

The easiest course for politicians has been to speak earnestly about long-term targets while avoiding difficult short-term steps that might cost political support. That way, politicians can be rewarded for their apparent virtue without imperilling their re-election prospects. That way has been the Canadian political way since climate change was first discussed at home and abroad.

CHAPTER TWO

Canada's Do-Nothing Strategy

The word came to the Canadian delegation from the prime minister himself. Forget the instructions the delegation had received as it departed in December 1997 for international negotiations on climate change in Kyoto. Forget the agreement negotiated with the provinces a few weeks earlier, for Ottawa had already unilaterally countermanded it. Now, Jean Chrétien sent fresh instructions: Canada must accept even more difficult targets for reducing greenhouse gas emissions.

Officials were incredulous. Provincial officials were steaming. Natural Resources Minister Ralph Goodale, co-chair of the Canadian delegation and a climate change skeptic, phoned a cabinet colleague in Ottawa and muttered the *r* word, *resignation*, a position from which the political lifer later retreated. But he did ask officials: Do we really have to follow this instruction? To which the answer, in Canada's prime ministerial system of government, was obvious. Without consulting cabinet, Chrétien had changed Canada's negotiating position.

From that moment on, guided almost entirely by political optics, Chrétien set Canada on a climate change policy course whereby the government accepted onerous obligations without knowing how to fulfill them. Every day from Chrétien's fateful call to Kyoto to the end of his years as prime minister, and every day since he left office, Canada has moved steadily farther from meeting its Kyoto targets, compiling the worst climate change record of any major country that signed the Kyoto Protocol on global warming. Even the United States, a country that eventually refused to participate in the treaty, had slower growth of greenhouse gas emissions than Canada, as shown below.

Commitments and Actual GHG Emissions of Signatories to the Kyoto Protocol, 2005

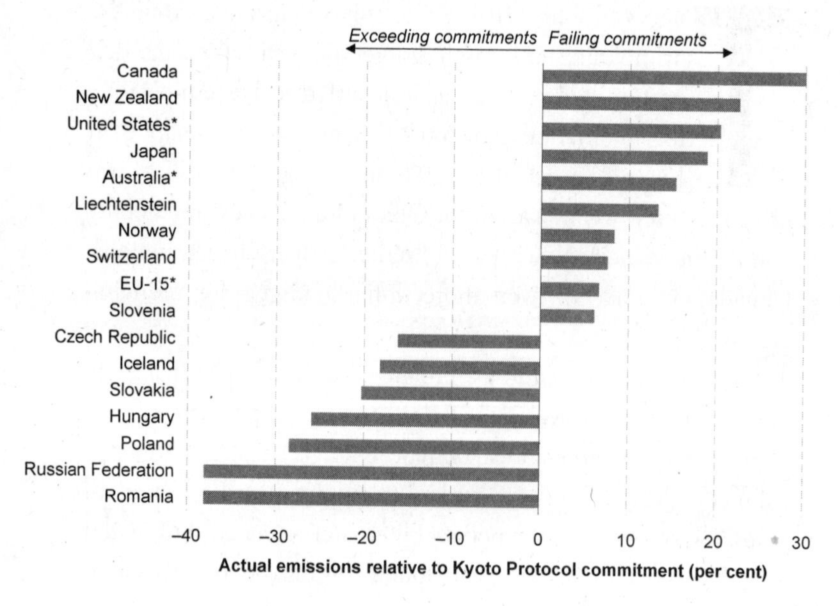

Source: Data from "Key GHG Data," United Nations Framework Convention on Climate Change, 2005.

* The United States and Australia have not ratified the Kyoto Protocol. The EU-15 have agreed to redistribute targets among themselves. Some economies in transition countries have baseline emissions other than 1990. Emissions and commitments exclude land use change and forestry.

Canada would always have struggled to be a vanguard nation in reducing GHGs, given its population growth, per capita wealth, robust economic growth, and active and powerful oil and gas sector, not to mention the automobile's significant role in the country, a cold climate, sprawling cities, and a federal structure of government. No prominent Canadian politician had consistently espoused the cause of climate change before Kyoto; no election campaign had featured the issue; no climate-induced catastrophe had awakened the public, although the ice storm that struck Quebec and Ontario in 1998 and the Saguenay floods in 1996 had started at least a few people wondering if something strange was happening to the weather. Considerable difficulty marked the negotiations between the federal and provincial governments that had finally produced a fragile agreement on Canada's negotiating goals three weeks before the conference began. The Kyoto negotiations, it was hoped, would result in countries committing themselves to specific GHG reduction targets for the end of the treaty's first phase, 2008–12. Once targets were identified, countries would then negotiate the range of acceptable means for achieving the targets and the verification procedures to ensure that countries did not cheat.

The federal-provincial agreement, negotiated at a meeting presided over by Goodale in Regina in November 1997, called for Canada to scale back its GHG emissions to 1990 levels by 2008–12. That target would have been very hard to meet, since between 1990 and 1997, emissions had already risen 13 per cent. They were still climbing at slightly less than 2 per cent a year. Canada therefore needed to stop that yearly increase as fast as possible, then reverse the trend. Each year of delay would mean a widening gap between commitment and reality. The faster Canada acted, the better its chances would be of achieving its Kyoto target. Nothing governments had said or done about reducing GHGs before

Kyoto, however, suggested that the federal-provincial target of stabilizing emissions was attainable. Governments before Kyoto had advanced only hesitant, milquetoast voluntary measures that had clearly been failing, as rising emissions demonstrated.

That the two levels of government had agreed upon a target, sort of, was what counted. "Sort of" because Quebec, with its abundant hydroelectricity, had wished Canada to go beyond just stabilizing its emissions. Alberta had remained prickly about anything that might injure its immensely wealthy and powerful oil and gas industry, and had let it be known that if the United States did not leave Kyoto with GHG reduction targets, or later decided not to ratify the treaty, Canada should abandon the Kyoto process. So great was provincial nervousness, and such was provincial constitutional control over natural resources, that Alberta, Ontario, and Quebec decided to send observers to the Kyoto negotiations to advise, warn, and watch over the federal negotiators.

U.S. president Bill Clinton and his environmentally conscious vice-president, Al Gore, had begun talking about global warming soon after their election in 1992. They had been pilloried for their interest in such tree-hugging sentimentalism by the oil and gas lobby, the coal industry, the automobile manufacturers, the U.S. Chamber of Commerce, other business organizations, and, of course, the Republican Party. The administration nonetheless persevered in the face of this concerted opposition and entered the Kyoto negotiations with a plan that the United States would reduce emissions to 1990 levels by 2008–12, the same position as Canada's. More ambitious than the North American position, however, were those being adopted by European countries, some of which were willing to reduce emissions by more than 10 per cent from 1990 levels.

In Ottawa, Jean Chrétien gave his initial instructions to Canadian negotiators: Canada must stay slightly ahead of whatever commitment the Americans might make. But not too far ahead, because the Canadian and American economies were so entwined that the smaller country could not afford – or so it was then thought – to accept a heavier burden than the United States.

A genuine negotiation is a place where fixed positions go to be modified. As Kyoto loomed, the U.S. administration seemed eager to bridge the gap between North America and Europe. President Clinton told Chrétien he hoped Canada would shift from its initial position, implying that perhaps the Americans might, too. The Chrétien government, reading these signals and entreaties, tried to anticipate how to jump just a little ahead of where the Americans might land. Since the United States might go from zero to perhaps a reduction of 1 or 2 per cent, Chrétien reckoned 3 per cent would be a safe political landing spot for Canada. The United States would never move that far, because Democratic and Republican members of the U.S. Senate, the body that would eventually have to ratify the Kyoto treaty, had already pronounced dead on arrival any treaty with the Clinton-Gore stabilization targets. Therefore, on the eve of the delegation's departure for Kyoto, the Chrétien cabinet announced that Canada would go to the conference prepared to accept an emissions cut of 3 per cent from 1990 levels by 2008–12. No serious economic work had been done to analyze how this new commitment could be met, but it would keep Canada slightly ahead of, and therefore more virtuous than, the United States, an important objective for the Chrétien government.

All hell broke loose in certain provincial capitals, especially but not exclusively in Alberta. Alberta's ministers fired off complaints to federal officials. Ontario's minister climbed on the plane for Kyoto and began demanding to know what on earth Chrétien was doing. Two factors made things even worse. All provinces were already furious with the Chrétien government for cuts to provincial transfers in the fight against the federal deficit. Second, the large provinces – British Columbia, Alberta, Ontario, and Quebec – were all governed by other than Liberal parties (not that shared political affiliation necessarily means friendship in federal-provincial relations). A federal government they had already accused of bad faith now again, after months of negotiating with the provinces, had demonstrated its untrustworthiness. Without forewarning, the Chrétien government had shattered the federal-provincial agreement at Regina of stabilizing emissions at 1990 levels and had accepted a heavier burden for Canada. If provincial representatives had been wary about federal intentions at their pre-Kyoto meeting in Regina, their wariness hardened into implacable suspicion before and at Kyoto. Suspicion would become hostility when they discovered Ottawa's next move.

IN KYOTO

Much of the Kyoto negotiations centred on what measures would be recognized as emissions reductions. A series of mini-negotiations on specific topics ensued that featured the big players such as the United States, the European Union, and Japan. Under the experienced leadership of Ambassador Paul Heinbecker, Canada usually managed to squeeze into these sessions. Natural Resources Minister Ralph Goodale and Environment Minister Christine Stewart formally led the delegation, but Heinbecker was the principal negotiator. He had been recruited from Foreign Affairs as a foreign policy professional, having served as deputy head of mission in Washington and ambassador to West Germany. He also could act as a referee in the internecine wars between Natural Resources and Environment, which were always at each other's throats. Sleep would be at a premium as negotiators wrestled with complicated issues, knowing that each country's emissions targets would grab the most attention.

Al Gore arrived to lead his country's delegation, signalling the importance the administration placed on climate change. The vice-president was also eyeing the presidency. He had written and spoken extensively about environmental issues, so this was his turf. He wanted the United States to play an international leadership role – and he wanted to be seen as having personally played such a role – even if his country could not go as far as the Europeans. The collapse of Communism had led to the closing of filthy old factories in the eastern half of Europe, a happy coincidence that meant that countries such as Germany and Russia could do nothing and still watch their emissions fall as the old plants closed. Other European countries were small or poor or had declining or stable populations, all factors limiting emissions growth. The United States was large, rich, and growing. Under these circumstances, it could not match the European targets, but Gore nonetheless felt his country could do better. He therefore suddenly announced that the United States, the world's largest GHG emitter, would reduce its emissions by an astonishing 7 per cent from 1990 levels.

Gore's declaration stunned the Canadian delegation and left Canada's 3 per cent reduction target looking feeble by comparison. Chrétien's negotiating instructions had been to stay slightly ahead of the Americans, assuming they would not move far, if at all, from zero. Now, in the glare of international publicity, Gore had upped the U.S. ante way beyond anything Chrétien had imagined. Canada had been blindsided. Frantic phone calls ensued between Kyoto and the Prime Minister's Office.

What to do? Provincial representatives, except for those from Quebec, were already furious at Ottawa's pre-Kyoto duplicity. Now Canadian ministers were being told by their civil servants that Canada need not go farther, even though Minister Stewart remained keen that Canada should stretch its target. What might the Americans do to Canada if they accepted a heavier burden for GHG emissions? Some sort of trade retaliation? Was Al Gore serious? Was this grandstanding for his presidential ambitions? Could his administration deliver in the Senate? American governments often used Senate ratification for their own purposes, demanding new concessions in international negotiations to increase the likelihood of congressional approval or, as in this case, taking positions while knowing the chances of approval back home were uncertain at best. Inadequate economic analysis had been done of the measures needed for Canada to meet a zero GHG target, let alone a 3 per cent reduction. Nothing at all had been prepared for an 8 to 10 per cent drop, the reduction required to meet the prime minister's instructions of remaining slightly ahead of the Americans.

The Canadians pulled back, ever so slightly. Eight per cent was too much, but how about 6 per cent? At 6 per cent, Canada would remain close to the Americans. Canada would not be morally superior, but perhaps it would not be economically disadvantaged either. Further negotiations about the means of reducing emissions lay ahead. Canada could always alter course if the U.S. Senate rejected the treaty. Nothing was carved in stone until ratification. But the political optics of the United States accepting a more onerous burden than Canada would be bad in Canada, and perhaps overseas.

The *Toronto Star*, environmentalists, a handful of Liberal MPs, and other self-appointed custodians of Canadian virtue would be outraged. True, Australia, another large energy producer, wanted to let emissions rise 8 per cent higher than 1990 levels by 2008–12. Norway, a country with an enviable record of constructive international engagement but also an oil and gas producer, was arguing for and winning a 3 per cent rise. These countries, however, could define their self-interest unencumbered by the political need to beat the Americans, whereas Canada, at least in Jean Chrétien's reading, could not.

And so, after many messages between Ottawa and Kyoto, Canada offered a 6 per cent reduction of GHG emissions from 1990 levels by 2008–12, without the foggiest idea how this commitment could be achieved, and over the fierce objections of most provinces. As a commissioner of the environment and sustainable development, Johanne Gélinas, later remarked, "With regards to the specific target, we found that little economic analysis was completed, and the government was unable to provide evidence of detailed social, environmental or risk analyses."

As a sign of things to come, Heinbecker dutifully followed a final instruction. Having accepted a burden without understanding its implications, the government began searching to close the gap between commitment and capability. At Kyoto's final session, when most of the delegates were tired and not paying much attention, Heinbecker rose and read for the record that Canada, in accepting the 6 per cent reduction target, would also be seeking credits for "clean energy" exports to the United States in the form of electricity and natural gas. No other country had signalled a willingness to accept this Canadian demand. Canadian officials were almost certain the demand would never be allowed. They were right.

PAVED WITH GOOD INTENTIONS

Chrétien's Kyoto instruction – to accept a minus-6 target for Canada – illustrated the Canadian propensity for talking grandly abroad while doing little at home to combat climate change. The pattern of making impressive international commitments without knowing how they could be kept started before Chrétien became prime minister and never varied until the Liberals lost power. Just as an "emissions gap" yawned between Canada's greenhouse gas reduction targets and actual emissions, a political hot air gap ballooned between government promises and delivery. Governments produced "action plans" and "green plans," with exhortations, good intentions, and elaborate possibilities for action, all of them masking what a close reading of the texts revealed: Canada's targets would not be met.

The desire for admiration, coupled with the instinct to compensate for modest capabilities with vaulting international ambitions, fuelled Canadian governments' early interest in global warming. Canadians, after all, were among the first climate change prophets, even if for a long time they lacked an audience in their own land.

When Brian Mulroney became prime minister in September 1984, climate change as a subject was unknown to the general public and remained inside the eccentric domain of scattered climatologists and environmentalists. A few scientists in the federal government and Canadian universities belonged to a small band of international worriers whose writings circulated within the rarefied world of conferences and academic journals.

In November 1984, Environment Canada sponsored a climate change conference in Toronto at which speakers warned of major disruptions. Significant warming will occur by the middle of the next century, cautioned the research chief of the Canadian Atmospheric Environment Service. Toronto's weather could become like that of Nashville, Tennessee, claimed the head of the Canadian Climate Centre. Opinions were divided, however, on the effects of climate change. The scientist Kenneth Hare suggested that Canadians "have survived a terrible climate for millennia," so it was hard to imagine that, on balance, a warmer climate could be worse. Some speakers predicted lower heating bills, which sounded good, while others warned of lower water levels to produce hydroelectricity, which sounded bad. Prescient predictions were offered, including easier navigation in the Arctic that would open the Northwest Passage to shipping, threats to the boreal forests, increasing drought conditions in parts of western Canada, a longer growing season, and dangers of coastal flooding, including to parts of Prince Edward Island.

Scientists shared this information, but only within their silos. So did the federal Environment Department, also called Environment Canada. The department had been created in 1972 as a kind of grab bag into which were stuffed responsibilities for forestry, fisheries, parks, and environmental studies; some of these were later removed and placed in separate departments. Environment, however, lacked tough regulatory authority and, most critically, lacked allies anywhere else in the Ottawa bureaucracy. From its creation, the Environment Department operated like a lost lamb in Ottawa circled by natural predators. Departments that worked with the energy sector – Energy, Natural Resources, and Industry – considered Environment a job-killing engine. Agriculture and Fisheries (once it became a stand-alone department) represented interests that thought Environment plotted to make their lives more difficult. Finance and Treasury Board viewed Environment as a nest of big spenders. All the political pressure in Ottawa, regardless of the party in power, to promote growth, accelerate regional development, and, sporadically, to restrain government spending left Environment without an internal constituency. The department's isolation reduced its influence on overall government policy, a weakness that persisted throughout the years of political hot air about climate change.

In 1985, when the department released a 35-page report on the state of Canada's environment, the paper found few listeners beyond those already attuned to the challenges of environmental protection. The developing departmental consensus about the science of climate change coloured a handful of pages in the report. Within the next 50 to 75 years, the report argued, Canada would start experiencing warmer weather, with increases ranging from 3 degrees in southern Canada to 10 degrees in the Arctic. Warmer weather, it continued, could cause prairie drought, more plant growth, increased agricultural activity in cooler and moister parts of Canada, a drop in Great Lakes water levels, less Arctic ice, higher sea levels, more and bigger forest fires, and widespread pest hazards.

Domestically, climate change remained a boutique issue, inside the government and beyond. The theory that government can effectively focus on only one issue was on display. The Mulroney government had identified an air quality issue, acid rain, as important and had mobilized efforts to improve Canada's performance and prod the United States into a cross-border agreement. Eventually an acid rain agreement was reached, but many of the important lessons learned about how to reduce emissions that caused acid rain, notably the role of mandatory caps on emissions and trading of credits, were later forgotten when it became time to design policies for lowering GHGs. Instead of action, Canadians got the creation of a national task force on environment and the economy under the Canadian Council of Resource and Environment Ministers, which duly published the first in a long series of reports whose influence on policy ranged from marginal to nil.

ENTER "SUSTAINABLE DEVELOPMENT"

An external jolt in 1987 injected political energy into all environmental issues, including climate change, with the publication of the report of the World Commission on Environment and Development, chaired by former Norwegian prime minister Gro Harlem Brundtland. Here the phrase "sustainable development" first burst from the cocoon of the academy into broad public use, with the commission's dire warning that the world, especially the industrialized part, was consuming too many resources for the planet's environmental health. "Many parts of the world are caught in a vicious downward spiral . . . making survival even more difficult and uncertain," *Our Common Future* warned. The 383-page report highlighted many environmental challenges and included a warning about climate change. "Energy efficiency can only buy time for the world to develop low-emission paths based on renewable resources such as hydro power, solar and wind power and biomass energy," it said.

The Brundtland report reverberated around the world. So did the 1988 report of the Intergovernmental Panel on Climate Change (IPCC), the first of a series of United Nations reports underscoring the perils of climate change and the urgency of remedial measures. Dissident scientists and climate change skeptics attacked the report as methodologically unsound. (A dwindling number of these types still attack the IPCC findings, which have since been updated and refined but not fundamentally changed.) The first report said world temperatures would rise by an average of 3 degrees by the end of the 21st century. The panel concluded, "There is a consensus that the risks associated with inaction on greenhouse gas emissions are too great to wait upon the results of further research before tangible steps are taken to address the problem." At the very least, the emerging science of global warming suggested governments should adopt the "precautionary principle," whereby even if the planet were warming more slowly than the IPCC believed, prudence still suggested reducing GHGs.

In Canada, the two reports confirmed what scientists and policy-makers in the Environment Department had been arguing – to little avail – in government circles. They also accelerated an upsurge in environmental interest that focused on two issues:

acid rain and depletion of the ozone layer. The upsurge did not long endure (such was the fate of environmental issues that they sometimes flowed but usually ebbed in public concern, unlike health care, the economy, and tax rates), but while it did, the Mulroney government began to pay attention to other environmental challenges as the 1988 election drew near.

In that year, the government played host to the World Conference on the Changing Atmosphere, the first major scientific conference on global climate change. The gathering was impressive for the numbers it attracted, important for the subject under discussion, and significant for what it illustrated at that early date about Canada and climate change. That Canada played host to the conference underlined the determination of the Canadian government to be a major international voice on the climate change stage. The Mulroney government committed the country to a 20 per cent GHG emissions reduction from 1988 levels by 2005, beginning the pattern of making grand international pledges without knowing how they could be achieved. Later that year, still desirous of playing a leading international role, Mulroney pushed to have climate change included on the agenda of the annual G7 Summit of the seven leading industrial countries.

If Mulroney's initial pledge – the first by a Canadian government – had remained Canadian policy, the country's emissions gap would yawn even more widely today, closer to 50 per cent than 35 per cent. Mulroney, as Canadians came to understand, was given to grandiose rhetoric. The 20 per cent reduction reflected that hyperbole. Inflated rhetoric aside, Mulroney wanted Canada to take an international lead on climate change, even if upon further reflection he scaled back Canada's commitment. He wished his government to be remembered as a friend of the environment during that period when Canadians were expressing environmental interest.

THE GREEN PLAN

After the 1988 re-election campaign, dominated by debate over free trade with the United States, Mulroney ordered a major overhaul of environmental policy. A huge interdepartmental team reviewed existing policies, laws, and regulations, and in 1990 produced "Canada's Green Plan," the first of five initiatives emitted by Ottawa in the next 15 years that dealt in whole or in part with climate change. The Green Plan outlined the most ambitious environmental agenda ever presented by a federal government, with recommendations ranging from fisheries protection to extension of national parks, from higher standards and stricter controls over drinking water to elimination of toxic wastes.

Mulroney had attended a Commonwealth Summit in Malaysia as work proceeded on the Green Plan. There, leaders received a report underscoring that global warming was a man-made problem that damaged future development prospects and threatened coastal states. Fifteen pages from the 174-page Green Plan paper, published after his return, were devoted to global warming. The government accepted the IPCC finding that scientific evidence pointed to the pressing challenge of global warming. The Green Plan warned that climate change might produce coastal flooding, changes in fish populations, prairie droughts, less snowfall in southern Ontario, and loss of Great Lakes ice. Under a business-as-usual (BAU) scenario, the document acknowledged, Canada's GHG emissions would grow by 1.6 per cent yearly from 1990 to 2000, a 17 per cent total compound increase.

Although the Green Plan had been developed under other ministers, its implementation and espousal fell in 1989 to Lucien Bouchard, the former separatist whom Mulroney had lured into federal politics. Bouchard's appointment to the Environment Department signified a break from past Liberal and Conservative practice, whereby ministers without influence landed at

Environment. Bouchard was Mulroney's personal friend, a political catch, and a man convinced of his destiny to guide his province's and his recently adopted country's future. Bouchard immediately injected his energy, charisma, and incandescent rhetoric into the job. About climate change, he declared soon after his appointment, "We must find a policy for this country to fight this terrible enemy, this subversive enemy, this silent enemy, which is destroying the planet now."

Having accepted the scientific reality of global warming, and having identified some of the threats to Canada, the government properly asked: What should be done? The Conservatives' answers, or rather non-answers, are worth noting, for they illustrate the gap between rhetoric and reality that plagued years of failed policies.

The Conservatives' Green Plan policies emphasized public education, government exhortation, and energy conservation by consumers. Nothing would be mandatory. No sacrifices or serious lifestyle changes were required. Climate change offered "opportunities," not losses. Everything was presented in an upbeat manner. Due account would be taken of "regional differences," the first government statement of the obvious – that GHG attenuation would be felt differently throughout a country with profound variations in methods of energy production and GHG levels. The government committed $85 million over six years – a piddling $14 million a year – to climate change research.

No economic instruments were suggested: taxes, market trading of carbon dioxide emissions credits, price changes, penalties, or rewards for appropriate behaviour. The absence of these instruments did not jibe with the Brundtland commission's recommendation that markets could be used to produce better environmental behaviour. The Mulroney government ignored these tools – just as subsequent Canadians governments largely did – to avoid fights with energy-producing provinces, industry, or unwilling consumers. The government greatly preferred mild, inoffensive, and largely ineffectual measures such as better labelling of consumer products, minimum efficiency codes, updated building standards, enhanced research and development, more tree planting, and a major public information campaign. Everything necessary was discarded; everything attractive but marginal was included. Not a word appeared about industrial emissions, despite the recognition that coal-fired electricity plants, oil and gas extraction and development, and automobiles contributed hugely to GHGs.

For the first time, but certainly not the last, the government admitted to the inadequacy of its measures while still insisting that Canada was leading the international battle against global warming. The document noted, "The measures outlined are, of themselves, unlikely to realize Canada's stabilization target. However, these initiatives will lay the foundation for achieving that objective."

What had that objective become? Mulroney's initial commitment to a 20 per cent reduction from 1988 levels by 2005 now became stabilizing GHG emissions at 1990 levels by 2000. The 20 per cent target had been approved by a Commons committee, dominated by Conservatives, but it proved too much for Canada's provinces, especially the oil-and-gas-producing ones, whose opposition provoked the first of many federal-provincial battles over Canada's climate change policy. The Mulroney government and the provinces shrank from emulating British prime minister Margaret Thatcher's embrace of the IPCC report and her pledge to slash Britain's GHGs by 30 per cent by 2005. Provincial energy ministers, gathered in the energy capital of Calgary, declared "premature" a 20 per cent commitment. At another meeting in Alberta, provincial ministers noted that a 20 per cent national cut would "cause significant dislocation and would require significant changes in life style." The first public evidence thus appeared of Alberta's and other provinces' objections to GHG reductions: economic woe, unwanted "life style" changes, unnecessarily speedy implementation.

The Mulroney government candidly explained its shift from a 20 per cent reduction to stabilizing emissions. We do not know how to achieve a 20 per cent cut, admitted Environment Minister Robert de Cotret at an international conference in 1990. That refreshingly frank admission, however, drew a blast from someone who many years later would widen the hot air gap between Canada's rhetoric and performance. "If your objective is to be one of the pack, as opposed to leading the trend, then Canada performed acceptably at the conference," said Liberal environment critic Paul Martin. "Mediocrity on the environment, however, is not good enough for Canada."

The new, more modest objective of stabilizing emissions at their 1990 level would have still required about a 16 per cent drop in GHG emissions by 2000. The Mulroney government never explained how to meet that target before the Conservatives left office in 1993. Nor did the government ever fulfill two other promises in the Green Plan: a discussion paper on using "economic instruments to achieve environmental objectives," and creation of an "inquiry into the environmental impact of electrical generation projects."

THE ROAD TO RIO

Inert at home, the Mulroney government remained active abroad. The Brundtland commission and IPCC reports had galvanized a United Nations process to focus international attention on climate change. The Mulroney government was determined to put Canada in the limelight, egged on by Maurice Strong, the Canadian member of the Brundtland commission. Strong subsequently became chairman of the meetings that culminated in the Earth Summit of 1992 in Rio de Janeiro. A large Canadian contingent of government officials and environmental groups descended on Rio, determined that Canada should be a prime mover, with the Europeans, in creating a treaty to reduce GHGs. Jean Charest, an energetic Mulroney minister with leadership ambitions, led the formal Canadian delegation. He had been a keen booster of the Green Plan in cabinet and across Canada, and wanted Canada to be seen abroad and at home as an avant-garde country opposed to climate change.

The Rio result produced the United Nations Framework Convention on Climate Change (UNFCCC), an international treaty aimed at reducing emissions. The UNFCCC nicely fit Canada's objectives. It required nothing concrete, relied essentially on voluntary measures, and therefore contained no obligations or penalties. It was an international treaty easily embraced by a country that loved multilateralism. The overall Rio target of stabilizing world GHG emissions at 1990 levels by 2000 was Canada's. It put Canada in the forefront of an international effort. Multilateral commitments, even if voluntary, might make it easier politically to sell difficult measures domestically. The treaty did produce a framework from which mandatory GHG targets could eventually be assigned to specific countries – a framework that would produce the Kyoto Protocol five years later. The UNFCCC targets, however, were voluntary. The government quickly ratified the treaty. The next year, Mulroney was gone from office. So were the Conservatives for another 13 years. Their actions never matched their rhetoric, a pattern the Liberals quickly adopted.

THE LIBERAL RED BOOK TO THE RESCUE

The Liberal Party, out of office for nine years, spent considerable effort drafting ideas for government. The leader, Jean Chrétien, was being called "yesterday's man." He had been around politics for three decades and was widely thought to be long on political instinct but short on policy ideas. Chrétien therefore asked his leadership rival, Paul Martin, and the Liberal research director, Chaviva Hošek, to tour the country and prepare a document on which the Liberals could base their 1993 election campaign and, if successful, their government. The ensuing "Red Book" did indeed become the Liberals' political guide, at least for campaign purposes. Asked what he would do about any policy area, Chrétien held up the Red Book and declared, "It's all in here."

The Red Book picked up the environmental zeitgeist of Brundtland and Rio. The Liberals borrowed the Conservatives' rhetoric that Canada should be an international leader in environmental protection. A chapter on "sustainable development" argued that "managing economic development and human growth without destroying the life-support systems of our planet demands of Canadians a fundamental shift in values and public policy." Liberals denounced "the gap between rhetoric and action under Conservative rule [that] has been most visible in the area of environmental assessment." Huge economic opportunities awaited Canadian companies, many of which "fortunately . . . are discovering that 'green economics' is the economics of efficiency. . . . Innovative solutions to domestic environmental problems can be marketed worldwide." Twenty-five per cent of all new government funding for research and development would be directed to technologies that "substantially reduce the harmful effects of industrial activity on the environment, or that specifically enhance the environment." These were ideas from Paul Martin,

Red Book co-author and Liberal environment critic; another Paul Martin, finance minister and prime minister, would emerge later.

On climate change, the Liberals grabbed a discarded Conservative target. Through energy efficiency and renewable energy use, the Liberals committed themselves to reduce carbon dioxide emissions by 20 per cent from 1988 levels by 2005 – the original Mulroney promise, made in 1988 but later eased in the face of provincial objections and the difficulty of hitting the target. The Liberals, five years and millions of tonnes of additional emissions later, promised to accomplish in 12 years what the Conservatives eventually realized could not be done in 17. The Liberals insisted the 20 per cent reduction target could be achieved "while maintaining a competitive economic base."

THE ALBERTA FACTOR

The economic base of the fossil fuel industry lies in Alberta. No industry emits more carbon dioxide than oil and gas, and the principal products of that industry – gasoline, diesel, and jet fuel – power the transportation sector that, in turn, produces about a quarter of all Canadian GHG emissions. No serious carbon dioxide plan could occur without major reductions in emissions from this industry, something the industry was determined to avoid. Happily for the industry, Chrétien appointed Anne McLellan as natural resources minister. A law professor from the University of Alberta, she was one of only four Liberals elected from the province. Although at first largely unknown to the industry, McLellan quickly became its cabinet protector and confessor. Her political job was to protect the interests of Alberta, and that meant the interests of the energy sector. She pursued that objective tenaciously, and the Alberta oil and gas industry admired her for it.

The fallout from the National Energy Program should never be forgotten in analyzing the Liberals' record on climate change. The NEP, a massive initiative of the Trudeau Liberals after the 1980 election to inject the government into all sectors of the energy industry, became Alberta's leading indictment against the arrogance and dangers of the Liberal Party.

A general economic slowdown hit economies across the Western world in the first two years of the 1980s. The downdraft was particularly severe in Canada. To this downdraft, the NEP added velocity. Thousands of Albertans lost a lot of money in the NEP period. Companies went bankrupt. Rigs moved out of the province. The NEP was being weakened when the Liberals lost power in 1984; by 1985, the Mulroney Conservatives buried its last vestiges. But the bitter memory of the NEP never died in Alberta, as Jean Chrétien understood.

It fell to Chrétien to placate Alberta as best he could after the ministerial architect of the NEP, Marc Lalonde, was shifted from Energy to Finance. Chrétien, the new energy minister in the Trudeau cabinet, spent a lot of time cracking jokes about his distant French-Canadian relatives in Alberta, schmoozing with oil and gas company executives, and easing some of the NEP's most injurious elements. When he became prime minister, Chrétien therefore knew, more than anyone in his cabinet, how Albertans – and not just energy industry executives – had hated the NEP, blamed the Liberals for it, and would not soon forget what they considered a premeditated attack on their province. Any energy or environment policy under Chrétien would be solicitous of the energy industry. McLellan's appointment testified to that desire for peace, which was underscored by her clout in cabinet. The Liberals' ground rules were established early: there could be a climate change policy, but it could not hurt Alberta.

Whenever the Liberals later appeared headed toward tough measures against the oil and gas industry, their intentions faded in the face of Alberta's fury.

Albertans understandably distrusted the Liberal government. Alberta had only one minister in a government dominated by Ontario MPs. Into the environment portfolio, Chrétien dropped Sheila Copps, a left-of-centre Liberal from Ontario, who began immediately playing up the Red Book's sustainable development commitments, including the promise of a 20 per cent drop in emissions from 1988 levels by 2005. Rumours emerged from Ottawa that the federal Liberals might be thinking about a carbon tax to raise the price of fuels in order to discourage consumption. The Red Book had mentioned nothing about green taxes, and every Canadian politician remembered Prime Minister Joe Clark's miserable fate after he promised to levy an 18-cent-a-gallon excise tax on gasoline in his 1979 budget. The unfounded rumours nonetheless persisted, each seized upon or encouraged by Preston Manning's new Reform Party, rooted in Alberta but spread across western Canada. No matter how many times Liberal ministers denied considering a carbon tax, Reformers replied that western Canadians had been lied to and duped by Liberals before.

Chrétien felt something more had to be done to spike the rumours. In the Canadian parliamentary system, with so much power concentrated in the hands of the prime minister, only a statement from him could kill the rumours. Therefore, Chrétien spoke to a convention of Calgary energy executives early in 1994 and squashed all possibility of a carbon tax. "Relax, relax," he said. "It's not on the table, and it will not be on the table." With those words, Chrétien removed for the duration of his government the possibility of emissions taxes that could have helped reduce carbon dioxide emissions. Civil servants who later considered recommending such taxes always retreated in the face of this prime ministerial diktat. They should have more vigorously "spoken truth to power," but they did not.

In a free-market economy, prices influence innovation and technological choices, depending on product and circumstance. Taxes, in turn, are one input that influences prices. By committing his government not to use taxes, Chrétien removed his government's most effective instrument to influence technology choice and behaviour. It would prove to be a decision of enormous consequence, impeding sensible action against emissions.

The Red Book commitment. Provincial concerns about unilateral federal action. Nervousness in Alberta. A corporal's guard of Liberal MPs demanding action. Natural Resources Minister Anne McLellan and Environment Minister Sheila Copps saying different things publicly and disagreeing constantly at cabinet. Business lobbyists warning of economic disaster from strong measures; environmental lobbyists warning of environmental disasters without them. International meetings in Geneva and Bonn on climate change, with another one scheduled for Kyoto at which willing countries would make commitments. By early 1995, the Chrétien government found itself pushed and pulled on climate change, and not just from inside Canada.

THE CLINTON PLAN

In Washington, President Bill Clinton's administration had produced a Climate Change Action Plan in 1993, a contrast to the position of President George H.W. Bush, who had insisted throughout his four years in office that climate change needed more study before the United States could commit itself to action. The Clinton plan offered a less demanding target than the one to which the Liberals were still nominally committed – the 20 per cent drop. It committed the United States to stabilizing emissions at 1990 levels by 2000.

Clinton's proposed means were voluntary. The government would spend $1.9 billion, and industry was expected to spend $60 billion, although his action plan did not spell out how or why. As in Canada, green taxes were ruled out. Environmental groups panned the action plan; industry expressed relief. Industrial lobby groups had warned that taxes or regulations, or both, would cripple the economy. Their arguments prevailed with the administration, just as they had in Congress.

Clinton and his vice-president, Al Gore, were determined, however, to make climate change a defining issue for their administration, and to press other world leaders to breathe life into the post-Rio negotiations toward an international treaty. Clinton, a master of the half-measure and a practitioner of "triangulation" (whereby he would position himself halfway between the Republican and Democratic positions on issues), was willing to spend some political energy on the climate change file, but not to the extent of risking confrontations with industry or incurring political unpopularity. His government had already been beaten back on health care reform; it would not risk another bruising (and likely losing) battle with industries as powerful as automobiles, oil and gas, coal, and agriculture, where farmers had not yet seen the advantages of biofuels and the potential risks to their livelihood of changing climate. Voluntarism coupled with exhortation was Clinton's preferred approach – the same one favoured in Canada, a cost-free non-solution to a real problem. Clinton began talking to other international leaders, including his good friend Chrétien, about how the world must make progress against climate change. The Clinton plan showed the president committed to action, even if not terribly seriously, whereas north of the border the government remained at sea.

CANADA'S COMPETING POLITICAL POWERS

Canadian constitutional arrangements complicated coherent action. Provinces jealously guarded their constitutional power over natural resources. For greater certainty, as lawyers say, that constitutional power was enshrined in the Charter of Rights and Freedoms negotiated by Trudeau and Chrétien with the provinces in 1980–81. Ottawa negotiated international treaties, but if the subject of these treaties fell in provincial jurisdiction, provinces could balk at implementation. A consensus would be needed between the federal and provincial governments before Canada could effectively negotiate, ratify, and implement an international climate change treaty, since many GHGs came from the oil and gas industries and provincial utility companies. NDP governments in British Columbia and Ontario and cities such as Toronto and Edmonton demanded mandatory regulations to reduce GHG emissions; Alberta insisted on voluntary measures.

With another post-Rio, pre-Kyoto conference looming, this one in Berlin, the country's environment and energy ministers struggled for consensus. In 1995, they produced a joint plan called the National Action Program on Climate Change, the first report on Canada's policies toward climate change since the Mulroney government's Green Plan in 1990. Compromise and inertia were splashed across every page. The Liberals' Red Book commitment of a 20 per cent reduction became the one Mulroney (and Clinton) had settled upon – stabilizing emissions at 1990 levels by 2000. Yet even stabilization would still require at least a 13 per cent reduction in the five years remaining before 2000.

The joint report did acknowledge the reality of climate change, a victory of sorts for those who wanted action against those who denied that humans represented the principal cause of global warming. "If climate change occurs to the extent predicted by current models, there will be a significant risk to the global environment, with potentially serious consequences for the health of the Canadian economy, particularly agriculture, forestry and fisheries," the report concluded. The report recognized that Canada was the world's second-largest per capita producer of GHG emissions.

Canada faced a choice between being "proactive" or "reactive." The country must be "proactive," the report declared, but the measures that followed were all voluntary, except for governments' commitments to reduce their own energy consumption. Again, carrots-without-sticks formulas emerged, as governments committed themselves to "educating Canadians," "promoting and celebrating successful and measured reductions," and "encouraging renewable energy."

Only two concrete measures were proposed, both of which were exposed in time as crashing failures. A national Voluntary Challenge and Registry was established, a clearing house for industries to report what they were doing about climate change. More ballyhooed still was the Canadian Industry Program for Energy Conservation, in which "specific energy targets will be developed for the major industrial sectors." Nine years later, the commissioner of the environment and sustainable development reported that this program had reduced emissions only .04 million tonnes.

Copps called the joint report a "disappointment." She took the Liberal Red Book promise seriously but discovered that many cabinet colleagues did not. As she later wrote, she ran up against an "anti-environment lobby that included the government's own all-powerful Department of Finance [and] Natural Resources and Industry [who] believed their clients believed in the status quo." She found morale in the department very low. "I remember one assistant deputy saying to me, 'Minister, I have outlived seven of your predecessors and I expect to outlive you. I have to get along in this town.'" Copps then supplied the translation: "Any proposed legislation had to have the blessing of Finance and Natural Resources, or it could not go ahead."

Copps seldom bit her tongue or adhered strictly to cabinet guidelines. At Berlin, she made a series of statements indicating that Ottawa would push for stronger action, thereby irritating provinces and deepening suspicions in Ottawa that the Environment Department and its minister were solo flyers, in cahoots with environmental non-government organizations to take Canada beyond where cabinet was prepared to go. At one point, Copps intercepted a phone call that led her to believe – correctly, as it turned out – that Anne McLellan was working within cabinet to have her recalled from Berlin. Certainly ministers were upset that she had apparently gone beyond her negotiating instructions in Berlin. Copps's statements further isolated her department, and her, within the closed bureaucratic circles of official Ottawa.

The continuing isolation of the department, the disagreements within the federal government, and the fragile state of federal-provincial relations over climate change would not disappear as Kyoto approached, even if by then Copps had moved to another portfolio. In her book *Worth Fighting For*, Copps looked back on the environment portfolio and concluded that "environmental interests were often trumped by commercial interests. Whatever issues came to the table, the environment minister found himself or herself in a minority. . . . As long as the cost of doing nothing was more attractive to business than doing something, the corporate lobby, especially the oil and gas lobby, would always convince the government that waiting was better than acting."

TOWARD ACTION PLAN 2000 ON CLIMATE CHANGE

After Kyoto in 1997, business could justify a strategy of denial and delay in part because the Chrétien government had no idea how to meet the obligations it had negotiated at Kyoto. To put matters politely, inadequate policy analysis had preceded the Kyoto negotiations, so that the Chrétien government, post-Kyoto, lacked concrete plans or credible policy analysis for reducing GHGs. Its changing positions had alienated the provinces during the negotiations. The government therefore faced at least some provinces that felt betrayed, a hostile business community, an empty policy cupboard, and, as always, a divided cabinet.

Had Canada left Kyoto with the target that had produced the fragile consensus at the Regina federal-provincial meeting, the federal government could have returned home and thought immediately about implementation measures. By unilaterally breaking that consensus in choosing its Kyoto commitment, Ottawa had to start over again, selling the reality of global warming, while trying to recreate a consensus around a new target. At a first ministers' meeting after Kyoto in early 1998, Chrétien tried to pour oil on troubled waters, as it were, musing aloud about other treaties that had not been ratified and agreeing to a joint federal-provincial study of implementation, a surefire recipe for delay.

Ottawa, from that first ministers' meeting forward, never stopped trying to find a new consensus but never succeeded. Starting over meant bringing all the interests back to a domestic negotiating table. Palavers among federal and provincial representatives and business were organized under the National Climate Change Process, with Ottawa and Alberta acting as co-leaders. Meetings were held by 16 "issue tables," involving 450 experts and 225 stakeholders, an exercise one participant described as an "Air Canada subsidy program," according to a paper by Professor Kathryn Harrison about the government's climate change policy.

The two-year, multimillion-dollar process of studies, consultations, and economic modelling finally culminated in a summary report in November 2000 on the economic and policy implications for Canada of the Kyoto Protocol. This report showed that GHG taxes of \$50 to \$150 per tonne of CO_2, or equivalent regulations that capped emissions and charged penalties for failure to comply (for business and households), would need to be implemented immediately in 2000 for Canada to have any hope of reducing domestic GHG emissions in line with its Kyoto commitment. That was the product of the massive consultation process. The next question was whether the government would implement anything of significance. It would not.

The futility of these consultations accorded with the Chrétien government's lack of genuine commitment and produced overall policy confusion. Very few ministers, apart from the environment minister, were even interested in the climate change file, and their reading of public opinion suggested that their indifference was mirrored by Canadians. This became evident in 2000, three years after the adoption of Kyoto, with the release of the third federal plan about climate change in 10 years. Action Plan 2000 on Climate Change indirectly acknowledged that Canada had negotiated its Kyoto commitment without any idea how to meet it. The paper bragged about the two years of consultation – "no other country has adopted such an open, inclusive and comprehensive process" – and declared that "this issue is one of the great environmental and economic challenges ever undertaken by Canada." Then the empty policy cupboard flew open. Three years after Kyoto, the best the Chrétien government could offer was a hoped-for 65-megatonne reduction – only one-third of what Canada needed to do to meet its Kyoto commitment.

The policies were the usual mix of exhortation and subsidy: consumer education, negotiations for fuel efficiency targets (an idea Ottawa first committed to in 1992), increasing ethanol production (requiring government subsidies), demonstration of best urban transportation technologies, research into fuel cells.

Everything was about "launching," "encouraging," and "developing"; nothing was about acting. Nothing, in particular, was demanded of the fossil fuel sector, whose emissions, the plan reported, were slated to increase by 65 per cent. For the utilities sector, whose emissions were going to increase by 24 per cent, the plan promised Ottawa would "initiate discussions with provinces and industry" and "work with interested jurisdictions."

The do-nothing approach to industry responded to the kind of argument the Business Council on National Issues had made to the 16 issue round tables when it warned of "substantial impairment of the Canadian economy" and blamed consumers more than producers for GHG emissions. Subsidies would be offered to municipalities for environmental improvements under the Green Municipal Investment Fund. No breakdown was given of how the 65-megatonne reduction would be achieved. Everything was voluntary and vague, despite a warning from Commissioner of the Environment and Sustainable Development Brian Emmett in his 1998 report that "the current approach, with its heavy emphasis on voluntary measures, is not sufficient to deal with the problem." Action Plan 2000 amounted to a road map to nowhere, which is where, predictably, it arrived.

INTERNATIONAL HORSE-TRADING

Meanwhile, another post-Kyoto exercise unfolded in a series of international gatherings designed to negotiate the rules by which countries could meet their targets. It was one thing for countries to accept targets, but another to sign an international agreement on what they would be permitted to do and how they would count their reductions. Three conferences were held. Before each, the Canadian negotiators' instructions precipitated rows inside cabinet between the Environment and Natural Resources departments. Natural Resources insisted that Canada negotiate credits for "carbon sinks," agricultural and forest land that absorbed carbon. These sinks, they argued, should count toward Canada's overall Kyoto target. The department also wanted countries such as Canada to be able to buy credits from developing countries, the theory being that the planet's health was indifferent to where reductions originated. Buying credits, of course, would mean fewer disruptions to the established ways of doing business in Canada. These arguments appealed to Chrétien, since they eased the economic pain and the political risks of complying with Canada's Kyoto's commitment.

A fundamental gap bedevilled these international conferences. On one side lay western European countries, led by Germany, whose coalition government included the Green Party. They sought mandatory rules, inflexible timetables, and penalties for non-compliance. They opposed carbon sinks as a sleight-of-hand and market-based systems for carbon dioxide reductions as too capitalist and difficult to enforce. They saw credits for transfers to developing countries as a way for developed countries to avoid changing their own practices. Of course, the collapse of the East German economy, and those of other former Communist countries in eastern Europe, made the general European position easy to understand. With polluting industries closing, meeting tough targets on strict deadlines would be much easier than anything required of energy-producing countries such as Canada.

Canada therefore lined up in 2000 with the United States, Norway, Australia, New Zealand, Iceland, and Japan, favouring international flexibility mechanisms such as carbon trading, sinks, and developing-country credits. The United States, then in the last months of the Clinton administration, was the intellectual leader of this group, fully engaged in the post-Kyoto negotiations, eager for success, but needing to sell the agreement back home on the basis of maximum flexibility for each country to design its own solutions.

Since Chrétien's decision at Kyoto to accept a minus-6 target, Canada had stayed close to the United States. Canada had been the Americans' best friend in the negotiations, taking ideas from American sources, supporting U.S. positions favouring flexibility of means, and trying to bridge gaps between the Clinton administration and the Europeans. Like the United States, Canada pushed hard for Brazil, India, and China to agree to participate in the next negotiations leading to post-2012 targets. Both Clinton and Chrétien felt the developing world needed help, and so pushed for buying credits from those countries. The dependence of the Canadian economy on the U.S. market and the geographic fact of sharing a continent suggested the smaller of the two countries should stick close to the larger. Alberta had always made it clear that Canada should never do anything about Kyoto if the United States did not ratify the agreement, and throughout the Clinton years a chorus of voices in the U.S. Senate kept warning that ratification would never happen, regardless of what the Clinton-Gore administration negotiated.

KYOTO DIES IN WASHINGTON

George W. Bush's victory changed Canadian calculations, but not immediately. The outgoing Clinton-Gore administration kept signalling that it might try to get the lame-duck Senate to ratify Kyoto. That forlorn hope never materialized. Christine Todd Whitman, the new Republican head of the Environmental Protection Agency, suggested her party did not oppose Kyoto per se, but had modifications in mind. The Bush administration, whose presidential candidate had not declared himself opposed to Kyoto during the election campaign, needed time to study the file and make proposals.

The Chrétien government paused, not wanting to get off on the wrong foot with the new administration, wondering what new ideas the Americans might provide. The negotiations about implementing Kyoto were almost over. All countries seemed more or less ready to ratify, although Canada continued to press for the inclusion of carbon sinks, a demand the Europeans still rejected.

In late 2001, George W. Bush, a Texan with deep personal and political connections to the fossil fuel industry, declared his opposition to Kyoto as harmful to the U.S. economy and pulled his country out of the talks. Australia, with a plus-8 Kyoto target, immediately followed the Bush lead. Alberta, the other western Canadian energy-producing provinces, and Ontario all demanded that Canada forget Kyoto and forge its own policy. Alberta premier Ralph Klein became so frustrated by Ottawa's refusal to ditch Kyoto that he asked if the federal government wanted Canadians to stop breathing. "Jesus, you know, I mean quit breathing? If all of us quit breathing, can you imagine how much carbon dioxide we could avoid sending into the atmosphere?" he told reporters. The business lobby redoubled its private efforts of persuasion and took out newspaper advertisements against ratification.

Kyoto doubters inside cabinet, including Natural Resources Minister Ralph Goodale, agreed that Canada should not go it alone in North America. Goodale had hustled to Washington six weeks after the inauguration to meet his new counterpart, Energy Secretary Spencer Abraham. He laid out the prospect of massive new developments in the Alberta oil sands to a cabinet secretary whose country needed new supplies from secure sources. "We have terrific business potential here – investment, jobs, growth – that can accrue to Canada's advantage. We want to maximize that," Goodale told reporters. As for the impact of the oil sands on GHGs, Goodale declared, "We are not convinced that the one necessarily must be sacrificed to the other. We want to pursue both these things simultaneously." Chrétien himself had delivered the same message to Bush at their first meeting, telling the president that the oil sands contained 300 billion barrels of crude oil reserves.

CHRÉTIEN'S LEGACY

The Chrétien government could have followed the American and Australian leads. Chrétien, however, remained committed to the climate change treaty, without indicating if or when Canada would ratify it. Increasingly, Chrétien saw climate change as a legacy issue that he wanted historians to place on the positive side of his leadership ledger. He would manoeuvre his government toward ratification and bequeath his successors the task of fulfilling commitments he had made.

Although often mocked inside Quebec, Chrétien viewed many national and international issues through the prism of his native province. The starting and end points of Chrétien's decision-making were conflated to one question: How would a decision fly in Quebec? Quebecers had been the most receptive Canadians to warnings about global warming. Quebec, with its massive hydro-electricity developments, produced the lowest per capita GHG emissions within Canada. Exploiting hydro required damming massive areas of Quebec, hardly an environmental plus, but the hydro produced a fraction of the emissions that came from burning oil, coal, or natural gas. Quebecers liked to think of themselves as environmental purists, even though their record in sewage treatment, toxic cleanup, and protection of lake water quality left them well down the Canadian charts. French-speaking Quebecers instinctively search for issues to display their moral superiority vis-à-vis the rest of Canada, just as Canadians seek opportunities to display their moral superiority vis-à-vis

Americans. Action against global warming allowed Quebecers to display their greater virtue in this one environmental field, as Chrétien understood.

The U.S. departure gave ammunition to Kyoto's critics in Canada. Canada found itself trapped. It had made a GHG reduction commitment without knowing how to meet it. Now, its continental partner, the United States, had left the negotiations. Australia, another large energy-producing country with which Canada had always maintained close relations, was gone. Alberta and other provinces clamoured for Canada to quit. Given these developments, before Canada could ratify Kyoto, Chrétien needed concessions that would reduce Canada's burden. He therefore instructed Canadian negotiators to press harder to win credit for sinks and to revive an old demand: that Canada receive credit for "clean energy" exports of natural gas and electricity to the United States. These exports showed Canada doing its bit for climate change. Canada demanded a 70-million-tonne credit for these exports, about a third of its overall Kyoto commitment, and arrogated that amount to itself.

Not a single country bought the Canadian argument. Kyoto clearly required GHGs to be counted at source. The other Kyoto signatories, especially the Europeans, considered the "clean energy" argument a brazen attempt to weasel out of a commitment. Canadian ministers insisted the gambit might work. Officials understood better and so informed ministers, who nonetheless peddled bogus arguments and inflated false hopes at home.

Canada's negotiators pressed on, achieving modest success. The U.S. departure made the Europeans more flexible, since the treaty had to be ratified by a set number of countries to come into force. As the ratification deadline approached, the treaty needed Japan, Canada, and Russia. Chrétien and his Japanese counterpart used their negotiating leverage, both insisting on sinks and credits for investing in GHG reduction projects in developing countries. Eventually, the Europeans conceded these points – but not the "clean energy" exports gambit.

The concessions did not satisfy Kyoto detractors, who by the summer of 2002 seemed a formidable array. Only two provinces, Quebec and Manitoba, now urged the federal government to ratify Kyoto. Mike Harris and Ralph Klein, Conservative premiers of Ontario and Alberta, remained fiercely opposed. Nova Scotia and Newfoundland, with offshore energy projects, slid into the Ontario-Alberta camp. Business remained unalterably hostile. Business-oriented cabinet ministers publicly expressed their concerns over Kyoto's negative economic impact. David Anderson, the environment minister, kept plugging Kyoto, even embarking on a cross-country speaking tour, but his prickly personality alienated cabinet colleagues. His department remained, as ever, isolated within the Ottawa bureaucracy.

PAUL MARTIN AND THE FINANCE DEPARTMENT

Paul Martin played an ambiguous role throughout the Kyoto debate. He had been the Liberals' environment critic in opposition, taking strong positions to limit GHGs. He had co-authored the party's 1993 Red Book, which called for a 20 per cent emissions reduction from 1988 levels by 2000. But as finance minister he said little about global warming. He seldom spoke on the issue in cabinet. His department did analyze the cost of climate change policies. These studies mostly demonstrated less economic disruption from strong measures than the business community or Alberta suggested, but for some reason they were never made public.

The Finance Department implacably opposed "green taxes," a position Martin faithfully defended in cabinet, because they smacked of using the tax system for social engineering objectives rather than raising revenue. Then, as now, the high priests of finance displayed an almost religious hostility to cluttering the tax system with specific measures aimed at producing behavioural change in consumers or taxpayers. These measures had grown, over the department's objections, during the social engineering years of the Trudeau governments, and since then the department had been steadfastly trying to eliminate them. Green taxes – to encourage, influence, or penalize certain decisions – represented a return to everything the department opposed in tax policy.

The Finance Department and its minister also apparently decided they did not want to "own" the climate change challenge, perhaps because it remained unclear just where the prime minister was headed – toward ratification or not? It was another example of civil servants not being resolute enough in "speaking truth to power." Finance intervened little in the internal climate change debate, except to squash talk of green taxes, a surprisingly passive and ultimately harmful rôle for the most powerful department in the federal government.

TWISTING CHRÉTIEN'S ARM: A QUESTIONABLE STRATEGY

Martin's ill-concealed leadership ambitions eventually became entwined with the climate change debate. After Martin's showdown with Chrétien greased his departure from cabinet, Martin's supporters intensified their search for anything to weaken Chrétien's position. As the Liberals prepared for their summer caucus in Chicoutimi in 2002, the pro-Martin forces gathered their strength to push Chrétien from the leadership. John Godfrey, a pro-Kyoto MP from Toronto, seized the Chicoutimi moment to circulate a petition demanding ratification. It secured support from 96 MPs (out of 172 Liberals) and 23 senators, including many pro-Martin MPs who by then represented the majority opinion in caucus, but who in this instance misunderstood Jean Chrétien. They assumed Chrétien would not ratify Kyoto, given cabinet divisions, business anger, and provincial opposition. They would embarrass Chrétien by signing the petition, because they could then use his forthcoming refusal to ratify Kyoto to illustrate, quite disingenuously, that the leader had once again given caucus the back of his hand.

Their misreading of his intentions delighted Chrétien. He had decided to ratify Kyoto, without of course knowing how Canada could meet its commitment. Indeed, he had been told by close advisers, including his chief policy adviser, Eddie Goldenberg, that the commitment could not be met. Chrétien had just not told anybody else. Ratification would bring together many valuable political objectives for Chrétien. It would differentiate Canada from the United States, a temptation not easily resisted by a Liberal government with George W. Bush in the White House. It would be popular in Quebec. It would form part of his legacy. It would be another in a series of signature decisions in his final years, including legalization of same-sex marriage, proposals to decriminalize marijuana, campaign finance reform, and keeping Canada out of the Iraq invasion. These measures would mock his tormentors' accusations that the old man had run out of political steam. He would leave in good time, with these accomplishments to his credit. When he did leave politics – the full force of the sponsorship scandal and the Gomery inquiry still lay in the future – the Liberals led in the polls, and his personal popularity had never been higher in Quebec.

Chrétien's closest advisers did not know his intentions when the prime minister departed after the Chicoutimi caucus for Johannesburg to participate in the World Summit on Sustainable Development, a conference marking 10 years since the Rio

Summit. The Prime Minister's Office even issued a statement that no announcement would be made at Johannesburg. There, however, Chrétien announced that Canada would indeed ratify Kyoto, having secured inclusion in the treaty of some credit for carbon sinks. Asked to explain his decision, Chrétien pointed to the caucus petition, among other reasons. In Canada, governments, not Parliament, ratify international treaties. The Chrétien declaration in Johannesburg, given a prime minister's control over cabinet and caucus, meant that Canada would formally ratify the treaty it had helped to negotiate five years before, a period during which emissions had grown and no effective policies had been developed to restrict them.

CHAPTER THREE

More Wasted Years of Talk

The Chrétien government, having ratified Kyoto, was still groping to understand how to achieve its target of bringing GHG emissions 6 per cent below 1990 levels by 2008–12. Five years had passed between the end of negotiations and ratification late in 2002, but nothing effective had been accomplished to close the gap between rhetoric and reality, between commitment and action. Civil servants had actually prepared a series of detailed possibilities, but ministers collectively had not come to grips with the climate change file, and the prime minister, having secured his place in history, was not about to plunge into the grubby details of implementation. Climate change policy would necessarily be a long-term business, and it was a complicated matter, too, not something to energize a prime minister on his way out of office.

Earlier, Chrétien had given instructions: there should be no steps toward implementation before ratification of Kyoto, just in case he decided not to ratify. The government's Action Plan 2000 had envisaged reductions amounting to only one-third of the Kyoto target, with only voluntary, vague measures that were destined to fall short of even that one-third. Chrétien's ratification announcement intensified the storm from business lobbies and recalcitrant provincial governments that now geared up for a final push to delay or block action. Paul Martin, free from ministerial solidarity, kept bemoaning the lack of a "plan" to achieve the Kyoto target.

Ottawa issued consultation papers that the angry provinces rejected. The papers reiterated some of Ottawa's oldest promises: no regulatory or tax measures (although civil servants had once again recommended privately in 2002 that a tax on carbon should be considered, only to have the idea shot down, once again, as politically unacceptable), and a reassurance to Alberta that no province would bear a disproportionate burden of GHG reductions. The former promise hemmed in the government; the latter promise was dubious. As Commissioner of the Environment and Sustainable Development Johanne Gélinas noted in 2006, "The impacts, costs and benefits of climate change will not be felt or shared equally by all Canadians – there will be economic, social and environmental winners and losers." Never did a Canadian government, before or after ratification, have the courage to explain this truth. They all preferred to peddle the illusion that every person and every region would benefit, and that no person or region would sacrifice. The consultation papers conceded through silence that none of the incentives the Chrétien government had given oil sands operators in 1995, including accelerated writeoffs, could be rescinded – not by a government that kept bragging to the world about the oil sands' potential. And, of course, no tax measures could be imposed on the consuming public.

THE CHRÉTIEN MANAGEMENT STYLE: AN INCONVENIENT TRUTH

With these restrictions limiting the government's room for policy manoeuvre, Chrétien instructed the Environment and Natural

Resources departments to develop yet another plan to allow Canada to meet its 1997 Kyoto target. Water and oil would be a symbolically accurate way of thinking of these two departments, and of their ability to produce a blended product. Environment thought of what economists call "externalities," the knock-on effects of certain economic decisions. Natural Resources thought of economic bottom lines, jobs, investment. Environment considered the consequences of producer and consumer behaviour; Natural Resources worried that those considerations might stifle important economic activity. Neither department trusted the science of the other. Their ministers often got along badly. In the post-Kyoto-ratification period, Natural Resources Minister Herb Dhaliwal and Environment Minister David Anderson, both from British Columbia, not-so-cordially disliked each other. In earlier days, as we have seen, mutual disdain marked relations between Anne McLellan and Sheila Copps. A cabinet committee, backed by a group of civil servants, was nominally in charge of coordinating the climate change file. In practice, the two ministers and their departments tried to hammer out compromises, sometimes with the guidance of the prime minister, but most often without.

In his book *The Way It Works*, Eddie Goldenberg, Jean Chrétien's longest-serving adviser, wrote of his boss's management style, "Chrétien had strong views on what a prime minister should and shouldn't do. He didn't have the time or inclination to be a details person, and was certain that wasn't the job of a prime minister. . . . Chrétien expected the PMO [Prime Minister's Office] to be a strategic player but he didn't expect it to micro-manage the whole government." This style effectively suited certain government files, especially those that remained within one department. If a department had a strong minister, or even a competent one, and that minister kept the government out of trouble, then

Chrétien's system worked well. Chrétien would outline broad directions, then expect ministers to carry out policies and seek guidance only if problems arose. Decisions got made quickly, the government stayed out of the headlines, the prime minister had time to focus on larger matters rather than micromanaging files. As minister and as prime minister, Chrétien did not bother with the details of long submissions. He prided himself on being a short-memo, clean-desk kind of leader. Two-hour cabinet meetings that started and ended on time astonished those who had endured the more discursive discussions under Prime Ministers Pierre Trudeau and John Turner.

Chrétien reserved a few very large issues for himself, especially those related to national unity and Quebec. For example, on something as consequential for his government (and the country) as the Clarity Act, Chrétien worked with one minister, Stéphane Dion, and then brought along cabinet, many of whose members had grave reservations about the wisdom of the legislation. The decision not to participate in the Iraq invasion was taken by Chrétien after listening to cabinet, but he informed only two ministers of his decision in advance.

Climate change, by contrast, was ill-suited to Chrétien's management style, with the predictable result that the gap between commitment and delivery widened throughout his years in power. Any climate change policy would be complicated, with so many interests at stake, including at least two opposed ministers and their departments, two levels of government, and a myriad of policy options. Climate change was also a new policy area. No previous Canadian government had attempted to reduce GHG emissions, so past practice could offer no guidance about what had worked or failed. Canada was trying to promote growth in energy production while simultaneously reducing emissions.

Only the most determined leader, one who identified the issue as an absolute top priority, could stickhandle the government toward a serious climate change policy. In Chrétien's government, the leader did not help carry the puck toward the net but rather sketched plays for the team on the chalkboard, then left the arena, returning to check out developments a while later. What climate change needed, given the internal government rivalries, was a prime minister who hammered decisions together, drove the bureaucracy, and expended political capital to persuade the public not just of the merits of the general direction of policy but of the sometimes painful measures the policy demanded.

Chrétien did the opposite. He intervened intermittently on the file, wrongly assumed ministers would hammer something out, delegated responsibility to staff, and expended very little political capital explaining why global warming represented a threat to Canada, let alone why Canadians would – not might, but would – have to change some of their ways to cope with the threat. He played to the international gallery about climate change, and he appealed to the muse of history to judge his years in office, but he did not indulge in the drearier work of making sure his policy succeeded.

"JUST ABOUT RIGHT"

Those who wanted a vigorous climate change policy pointed to Canadian polls to buttress their ambitions. During the Kyoto negotiations, an Ekos Research poll showed 67 per cent of respondents prepared to make adjustments, including a higher personal cost of living, if they believed that cost would limit or eliminate climate change. Pro-Kyoto boosters were delighted, except for the finding that only 43 per cent of respondents actually knew that human activity caused global warming. An Angus Reid Group survey after Kyoto found 9 out of 10 Canadians believed the agreement to be "just about right." Twenty-two per cent of respondents to this 1998 survey said global warming ranked among the top two international issues, up from 6 per cent in 1996. By 2001, a Decima Research poll commissioned by two environmental organizations showed that 9 out of 10 Canadians now believed human activity harmed the planet, and 93 per cent wanted Ottawa to do more to curb GHGs.

And yet, another Decima Research poll the next year, this one commissioned by the federal government, revealed the "shallowness of Canadians' environmental ethic." The poll found that "Canadians are not keen to make major shifts in their day-to-day life such as seeking out alternative energy sources." The poll discovered that Canadians were aware of global warming "at a rudimentary level, but most are not yet paying much attention to it, nor are they truly concerned about its implications for the country or for themselves." Canadians were confused. A significant number linked climate change to the ozone layer; one in four thought nuclear power – not a significant GHG emitter – contributed to global warming; 17 per cent thought solar activity or sunspots caused climate change.

Proponents and skeptics seized upon polling data to frame their arguments for or against government action. What counted most was the government's own intuitive reading of public opinion. That broad reading suggested that the public understood very little about the causes and effects of global warming, believed in an unfocused fashion that perhaps something should be done, but stoutly resisted lifestyle changes or higher taxes. Someone or somewhere else was at fault – other countries, industry, another part of Canada. Canadians did not see themselves or their lifestyles as responsible for climate change.

From the polls, the Liberals gathered the vital message that the appearance of action was more politically necessary than action itself, especially if that action entailed sacrifice. No Liberal returned from campaigns saying he or she had been elected on climate change, and none reckoned careers would rest on the issue in the next election. A cluster of MPs urged vigorous action, including economic measures. For many MPs, climate change was one of those notional, feel-good issues about which some electors felt strongly, but the rest did not. There were also a range of factors over which, if MPs were honest with themselves, governments had little control, whereas the ones over which they had influence, risked upsetting important constituencies.

A POLICY VACUUM ON CLIMATE CHANGE

The mixture of rhetorical good intentions and inadequate or inappropriate policies produced the vacuum that was, and is, Canada's approach to climate change. Even if the policy prescriptions had been better and the political will more resolute, Canada would still have faced an uphill battle to reduce GHG emissions.

To better understand Canada's challenge, think about the different factors influencing output of GHGs. Everything else being equal, emissions rise with population, per capita income, and economic growth, and with the prominence of energy-intensive (especially fossil-fuel-intensive) sectors in the economy. Against these pressures causing GHGs to increase, others can cause a decrease. Forestry and agriculture can be better managed to reduce GHG emissions. Energy efficiency can be improved. Consumers and industries can switch to fuels that produce lower GHGs. And emissions can be controlled so that they never reach the atmosphere.

The truth is that governments cannot do much about some of the key factors that drive up GHGs. No Canadian government favours lower immigration or a smaller population. None would pursue policies to drive down per capita incomes or economic growth. And although much has been made of switching to renewable sources of energy, Canadian governments have simultaneously promoted development of fossil fuels, especially from Alberta's oil sands, for which insatiable markets exist in Canada and the United States. Canadian governments can focus, and they have, on measures to reduce GHGs – that's ostensibly what Ottawa's five climate change policies since 1990 have attempted – but all of these have failed. Moreover, no targets or associated policies have adequately accounted for the unique demographic and economic characteristics that have driven Canada's emissions record ever upward.

Canada has a very high rate of population growth compared with most other developed countries. Statistics Canada reported in 2007 that Canada's population had grown by 5.4 per cent from 2001 to 2006, compared with 5 per cent growth for the United States, 3.1 per cent for Italy and France, 1.9 per cent for the United Kingdom, and no growth at all for Germany. As shown in this figure, Canada's population increased annually by 1.02 per cent between 1990 and 2005. Population growth in five of the other G7 countries was much slower; only the rate in the United States approximated Canada's. Since 2000, Canada's growth rate has exceeded even that of the United States, making Canada's population the fastest-growing of the G7 countries.

Canada's population growth results from a natural increase of about 0.4 per cent per year and net immigration of about 0.6 per cent. Immigration, then, is the driving force behind population growth. Canada's immigration relative to population

Population Growth in G7 Countries, 1990–2005

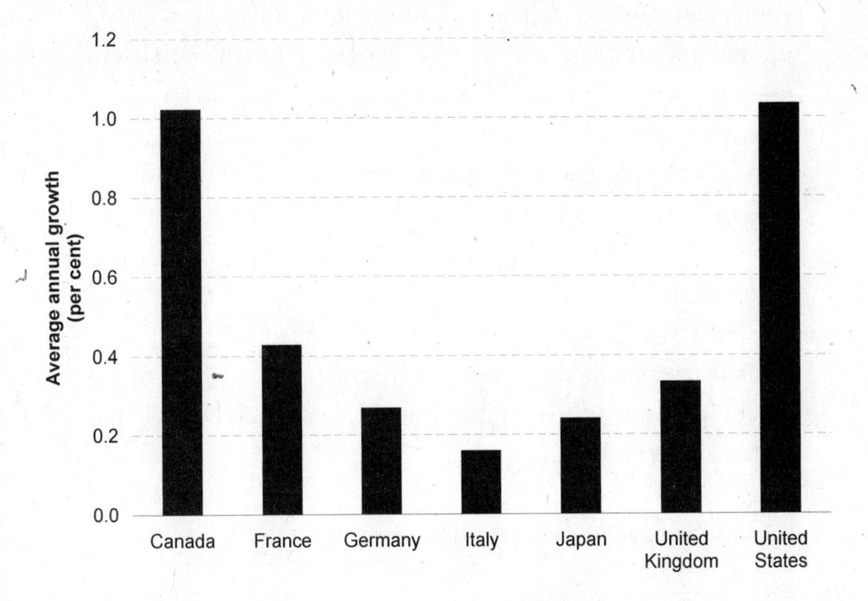

Source: Mid-year population estimated by the Population Division/DESA of the United Nations Secretariat.

has been the highest in the G7, about 17 per cent above that of the United States and far above that of Europe or Japan. If Canada's population, instead of growing at 1 per cent a year, had increased at the average European and Japanese rate of only 0.3 per cent, Canada's GHG emissions in 2005 would have been only 12 per cent higher than their 1990 levels, instead of 25 per cent higher.

Canada's population growth rate will remain above that of other advanced industrial countries for the foreseeable future. Canadian governments have no intention of lowering immigration intake; in fact, some politicians wish the intake to go higher. Canada's death rate is not expected to exceed its birth rate until

2025, so natural increases in population will also continue. Population growth, therefore, will continue to exert a strong upward pressure on emissions as long as our energy system includes energy forms and technologies that emit GHGs. Indeed, increasing population puts the single biggest upward pressure on Canada's greenhouse gas emissions.

Canada's economy has also grown impressively, with an annual growth rate of 2.8 per cent from 1990 to 2005, about equal to that of the United States, but significantly greater than that of the other G7 countries. When GDP is divided by population, Canada's per capita GDP growth has been faster than those of the other G7 countries, except the United Kingdom and the United States, as the next figure illustrates.

Rates of Economic Growth in G7 Countries, 1990–2005

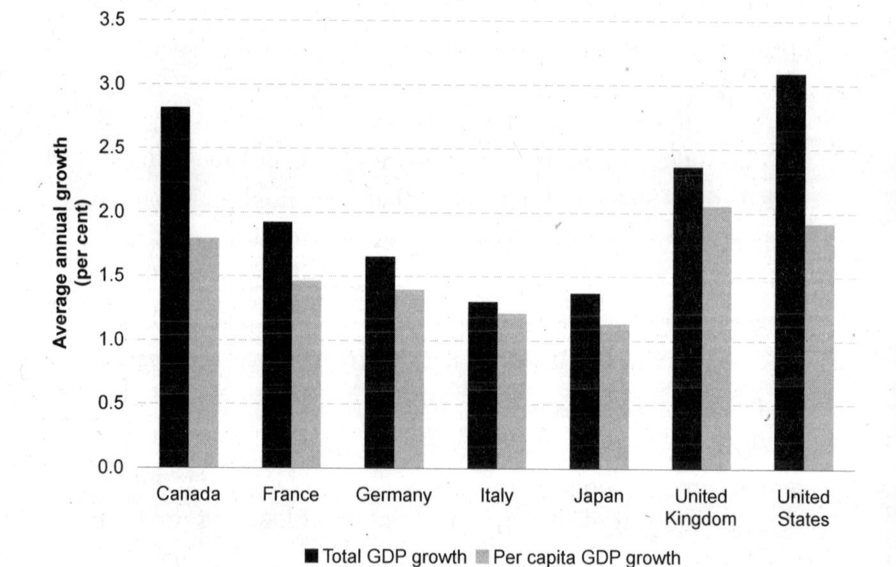

Source: World Bank.

All governments – everywhere, always – favour economic growth. Yet that growth will in and of itself drive up GHG emissions – unless these are offset by gains in energy efficiency, fuel switching, or emissions controls. And our economic growth is expected to continue to advance rapidly. A Conference Board of Canada projection suggests that Canada's GDP will increase at an average rate of 2.6 per cent through 2025. Canada's quickly expanding economy is second only to increasing population in explaining the country's increasing GHG emissions. If Canada's per capita economic growth rate had been 1.2 per cent annually, as in Europe and Japan, emissions would have grown by only 18 per cent from 1990 to 2005, instead of 25 per cent. If Canada's population *and* economic growth had grown at sluggish European rates, Canada's GHG emissions would have increased by only 6 per cent.

Add to the GHG challenge from population and economic growth several important structural aspects of the Canadian economy. With 174 billion barrels of proven oil reserves, and many more projected, Alberta's oil sands place Canada's reserves second in the world to those of Saudi Arabia. Limited commercial development of the oil sands started in 1967, but engineering challenges and high extraction costs constrained output for several decades. To get development under way, the Alberta government offered lucrative incentives – including the initial rock-bottom 1 per cent royalty rate – for oil sands development. Since the mid-1990s, technological advances, more tax incentives (including from Ottawa), and higher crude oil prices have stimulated significant expansion of oil sands operations. Daily oil sands production has tripled in volume since 1990. The National Energy Board forecasts continued rapid growth, with a quadrupling of output by 2020.

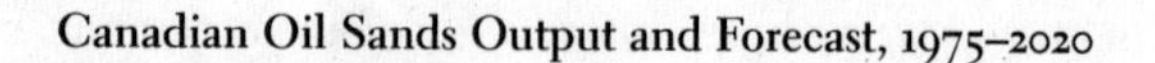

Canadian Oil Sands Output and Forecast, 1975–2020

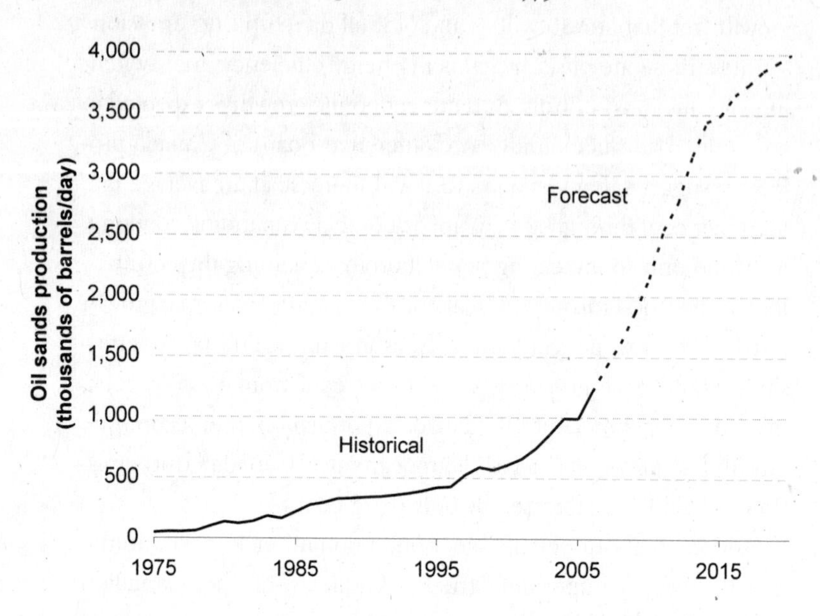

Source: Historical oil sands output from National Energy Board; forecast oil sands output from Canadian Association of Petroleum Producers.

Oil is extracted from sands either by surface mining (followed by washing to remove sand, silt, and clay) or by injecting steam into deeper deposits (in situ production) that frees the oil from the sand and drives it to the surface. The resulting heavy oil must be upgraded by adding hydrogen. Each step requires a significant amount of energy. The result is that oil sands production is much more energy-intensive, and therefore GHG-emissions-intensive, than conventional oil production. Production of a barrel of synthetic crude from the oil sands creates roughly 50 to 100 kilograms of CO_2 emissions, twice as much as a barrel of conventional light crude and five times as much as conventional light crude from offshore locations such as those in Norway and the United Kingdom.

Increases in oil sands production between 1990 and 2005 have pushed up Canada's GHG emissions by over 20 million tonnes. Without that growth, Canada's emissions would have climbed only 22 per cent, instead of 25 per cent. With a predicted quadrupling of oil sands production ahead, Canada's GHG emissions will increase quickly – unless serious measures are taken to prevent these anticipated emissions from reaching the atmosphere. Such measures have not been taken because for years now, Canadian politicians have continued to insist that the circle could be squared, with aggressive pursuit of oil sands development *and* lower GHG emissions. The Alberta government and the industry have counted on improved energy intensity – less energy used for each barrel produced – but these improvements have been and will be dwarfed by much higher output.

Growth in population, the economy, and the oil sands industry has driven Canada's GHGs ever higher over the past 15 years. Nothing suggests these trends will slacken, let alone be reversed. Indeed, government policies will actively promote growth in all three areas, thereby making it difficult for Canada's climate change policy to flatten emissions growth, let alone reduce GHGs. No other G8 country faces these pressures. Project Green, the Martin government's later plan to combat climate change, did not stretch the truth in observing that Canada's Kyoto Protocol target was "the most challenging among Kyoto signatories."

A CLIMATE CHANGE PLAN FOR CANADA!

Despite these formidable challenges, Canada under Jean Chrétien had made an international commitment, even if the country was moving farther from meeting it every day. Something new and better had to be crafted to move Canada along and convince foreign skeptics that Canada would be as good as its word.

The new policy, tugged into place by the Siamese twins of the Environment and Natural Resources departments in 2002, was called Climate Change Plan for Canada, and it repeated the now-familiar refrains about the hazards of global warming and Canada's determination to meet its Kyoto commitment. The plan acknowledged, rhetoric notwithstanding, that even if its policies were completely implemented, emissions would still be 70 million tonnes over Canada's obligations. The new policy offered possibilities for eliminating that gap, none of them credible, testifying again to the triumph of illusion over reality. The most improbable possibility was winning international support for the continuing claim of credit for up to 70 million tonnes from "clean energy" exports to the United States – the Chrétien government's forlorn hope that Paul Heinbecker had first put on the record in the final moments of the Kyoto conference, but which had been rejected ever since by all other signatories.

The new plan, like the three previous ones, relied extensively on exhortation. Canadians should join the One-Tonne Challenge. Each Canadian generated just over 5 tonnes of GHG emissions each year. These collectively amounted to a quarter of the country's total. If everyone lowered his or her GHG emissions by 1 tonne the country could lop 30 million tonnes from its emissions, or an eighth of its overall obligations. Canadians could meet the challenge by idling their cars less, buying more fuel-efficient vehicles, walking and cycling more, using ethanol-blended gasoline, lowering thermostats in winter, turning out lights, retrofitting older homes, and purchasing energy-efficient furnaces. Pamphlets would inform Canadians how to meet the challenge. The comedian Rick Mercer would be featured in television advertisements urging citizens to get with the program.

Predictably, Canadians would fail miserably to meet the One-Tonne Challenge, as subsequent studies showed. The television advertisements with Mercer ran from September 2002 to March 2003 at a cost of $17 million. When Decima Research later surveyed Canadians about the ads, most people said they had never seen them; those who had seen them did not know what Mercer was talking about. The Decima report concluded that the ads had been a waste of money.

Climate Change Plan for Canada credited Canadian governments with already having created programs that had slashed 80 million tonnes from the national target. This statement was political bafflegab, since almost half of these cuts had come not from changes in consumer or industry practices or from successful government programs but from credits Canada had negotiated at Kyoto for carbon sinks of forests and agricultural lands. Canadian geography and the determination of Canadian negotiators had produced half of these vaunted results, not government programs. The other half of the assumed cuts was pure conjecture; no credible analysis was provided in defence of this claim.

To secure the next 100 million tonnes of reductions, the plan offered pious hopes and good intentions, the theory being that 10 federal budgets remained before the expiration of the first phase of Kyoto in 2012. There would be time enough for more public transit, energy-efficient building codes, retrofitted buildings, renewable energy, better vehicle efficiency standards, and, of course, an accelerated take-up of the One-Tonne Challenge by ordinary Canadians.

THE LARGE FINAL EMITTERS

Climate Change Plan for Canada broke with other plans, however, in one important respect. It acknowledged what previously had been denied by silence: it admitted that no serious climate change policy could exempt the industrial sector, especially the so-called large final emitters (LFEs). The plan underscored that these companies were responsible for about half of Canada's GHG emissions. These companies, through their lobby associations, had always resisted mandatory emissions reductions. They preferred voluntarism, which had plainly been failing. They wanted only reductions in energy "intensity" – which sounded great until you understood that it simply meant continuing standard and sensible business practice to reduce costs. They never publicized the fact that if intensity falls more slowly than output rises, emissions would keep rising. What they feared most was a cap-and-trade system, of the kind being developed in Europe, whereby governments establish a carbon emissions cap or limit for an overall industry, gradually reduce that limit, and leave companies to change their procedures to reduce emissions or buy credits from others that have done more. This method was the most promising in lowering emissions, but there could be economic costs.

The Environment Department had pushed for such a system, but industry and its supporters in Natural Resources balked. Instead, the new plan called for negotiations with the LFEs toward individual "covenants," whereby each company would negotiate emissions permits with government. If companies fell below their allocations, they could sell the credit; if they were above, they could buy credits from other companies at home or overseas. Natural Resources Minister Ralph Goodale was certain he could negotiate this hybrid model with the LFEs. He failed before he left the portfolio, finding that negotiating covenants with dozens of companies was far too complicated.

Moreover, no sooner did the ink dry on the plan than the government resumed the familiar pattern of making concessions to industry. The plan had assigned a 55-million-tonne reduction target to the LFEs, but the government conceded – in the first of a string of compromises – that some companies might be allowed a longer time frame than 2012 to fulfill obligations, provided they made up the difference later.

Further concessions followed. Herb Dhaliwal, a very pro-business minister who replaced Goodale in a cabinet shuffle in 2002, got the LFE target reduced from 55 to 45 million tonnes, with the possibility of a further 6-million-tonne reduction if the companies invested in clean technology projects. He also agreed that the price for carbon emitted over negotiated company limits would be only $15 a tonne. This price, demanded by industry, was far below what independent analysts considered a sufficient incentive to drive substantial technological change.

PAUL MARTIN GETS HIS TURN

When Paul Martin became prime minister late in 2003, civil servants told him that existing policies, including Climate Change Plan for Canada, would not get Canada to its Kyoto target. One year later, a January 2005 government document leaked to the *Globe and Mail* said that "with current policy and programs, Canada is still going to be significantly off the Kyoto target." Government needs "more consideration of regulation and taxation to drive behavioural change and technology development and uptake." Moreover, the document warned (the warning was ignored) that the "voluntary approach and limited incentives are not sufficient to drive substantive change." Chrétien had left office, with Canada inside Kyoto but stuck with a plan that would not work.

Martin knew that already, since in the time between his leaving cabinet and his becoming leader, he had lamented the lack of a "plan," by inference criticizing the inadequacies of existing policies. But like his predecessor, he ruled out tax changes, avoided riling industry, and so forced Ottawa to rely even more heavily than before on exhortation and subsidy, the two approaches that had failed for so long.

Martin, elected with only a minority in 2004, made Stéphane Dion the environment minister, a recognition that he had erred in booting Dion, a loyal Chrétien supporter, from his first cabinet. For Dion, the environment portfolio represented a chance for political redemption. He plunged into the file with his customary seriousness and passion, having been one of the few ministers in cabinet who had backed David Anderson, his frequently beleaguered predecessor. He had the prime minister's blessing, a bulldozer of a deputy minister, and the zeal of a convert.

DION'S DILEMMA

What did Dion confront, as he supervised the preparation of yet another climate change plan? Canada's emissions in 2005 were 25 per cent higher than in 1990. The country's commitment at Kyoto had been to reduce them by 6 per cent by 2008–12. Canada was therefore already over 30 per cent above its Kyoto target, and heading upward at almost 2 per cent a year. At this rate, Canada would miss by almost 50 per cent by 2012! More astounding still, despite all the plans, announced programs, and rhetoric, of the $3.7 billion that governments had set aside for climate change, only $1.7 billion had actually been spent, or about $300 million a year since 1988.

Overall energy intensity had improved by 13 per cent since 1990 – a measly 1 per cent a year, an improvement overwhelmed by economic growth. This energy intensity achievement should never have been trumpeted by government, or by industry for that matter, since energy intensity had been improving throughout most of the previous century. Energy intensity actually improved more rapidly in the 1980s than in the 1990s and 2000s, because most businesses and householders in that decade assumed (wrongly, as it turned out) that energy prices would keep rising, as they had throughout the 1970s. Governments did not like the moral of that tale because of the political risks involved: prices, or beliefs about future prices, do influence behaviour. There is a corollary lesson governments have yet to learn: in future, the price of emissions should matter much more than the price of energy. We'll explain this lesson in more detail later.

After months of preparation, in April 2005, Dion presented Project Green, a policy of exhortation and subsidy on steroids. After the usual grandiose rhetoric about the importance of climate change, the plan unveiled $10 billion in spending that "could" get Canada to its Kyoto target. Canada needed a 270-million-tonne reduction; the LFES, however, were given a target of only 45 million tonnes. The automobile industry successfully avoided mandatory limits and negotiated instead a voluntary agreement to curb emissions by 5.3 million tonnes by 2010. Taken together, therefore, the major industrial and transportation sources of GHGS were asked to share less than 20 per cent of the national burden, which meant that by definition the entire plan was destined to fail. Huge subsidies came in two new programs: the Climate Change Fund, through which the government itself would buy credits, and the Partnership Fund, which it would use to invest in GHG-reducing projects initiated by other sectors. The famous but failing One-Tonne Challenge remained, as did subsidies for renewable energy.

Once again, despite the warning of the January secret document, the government had ruled out tax changes, differential pricing, market mechanisms (although it did suggest a possible cap-and-trade system in selected industrial sectors, which was promising), and stiff regulations, choosing to rely largely on an information and subsidy program. The Martin government fell before Project Green had any impact, but the sad history of the exhortation and subsidy model suggested it, too, would have failed, just more expensively than its four predecessors.

STEPHEN HARPER: COOL ON GLOBAL WARMING

Stephen Harper's Conservatives arrived in office early in 2006 knowing better what they opposed than what they favoured. They had devoted only two lines to global warming in their canonical campaign document, *Stand Up for Canada*, saying that the party did not like the Kyoto accord and would instead develop a "made-in-Canada" approach.

Harper, as leader of the Canadian Alliance, had railed against Kyoto, repeating business arguments that the "impact of reducing emissions on this scale would be enormous. . . . Canadians could be looking at 50 per cent increases in the cost of electricity, gasoline and heating, the loss of half a million jobs and economic costs of $40 billion." His "sensible alternative" was the usual stew of ineffective measures: encouraging energy efficiency, relying on intensity improvements, investing in research, and "monitoring the science to get a much better understanding of the extent of climate change and its causes." As an economist – and one based in Calgary, at that – Harper for a long time viewed climate change exclusively through the prism of what he perceived as the negative economic consequences of serious action, combined with his own skepticism about the science of climate change.

Harper had hammered together a merger of the Canadian Alliance and the Progressive Conservative Party to create the new Conservative Party. The Progressive Conservatives, when led by Jean Charest, had been moderately supportive of the Kyoto process, since after all Charest had been environment minister under Brian Mulroney and had led the Canadian delegation at the Rio conference. The Canadian Alliance and its precursor, the Reform Party, had been totally opposed to Kyoto, and deeply skeptical of the science of climate change.

THE MANNING LEGACY

As far back as November 1997, on the eve of the Kyoto conference, Reform leader Preston Manning spoke for 45 minutes in a "take-note" debate in the House of Commons, deploring all aspects of the government's climate change policy. Manning, like Harper later, decried the lack of analysis that would outline the "economic, sectoral, regional and taxpayer impacts of pursuing CO_2 emissions reduction targets." Manning was one of the few MPs deeply interested in science, and his reading led him to admit that "man's activities over the past two centuries, in particular the burning of hydrocarbons and the destruction of forest, have led to an increase of between one-quarter and one-third of atmospheric CO_2." But that same reading led him to conclude that the science was not conclusive, even though the majority of reputable scientists around the world were saying the contrary.

Manning, like other skeptics, chose to quote a variety of scientists who doubted the link between increased human-created CO_2 emissions and global warming. The Reform Party, therefore, was "sceptical about the alleged science behind the government's position." By largely dismissing the science of climate change, Manning and the Reform Party also dismissed serious action, a position that mirrored the positions of the fossil fuel industry and the Alberta government.

Kyoto, Manning asserted, would lead to "significant reductions in GNP over the next one or two decades." He complained that nowhere did the government answer the question, "Who pays?"

Unless it developed a plan that considered all the costs of Kyoto, and detailed how this plan would be financed, and which taxpayers and provinces would pay the bill, then the government should not sign anything at Kyoto.

Manning, as was often the case, presented an eloquent, reasoned argument, but because his alternative to Kyoto was so weak and his skepticism about the science so evident, Reform cast its lot in with not just the critics of Kyoto but the legions of doubters of the very existence of human-induced climate change. Manning gave the intellectual lead to his followers, not all of whom used his restrained language. Said Reform MP Jim Abbott, "This entire process is one of watching lemmings run. It is one of very questionable science." Kyoto, he continued, would likely "destroy parts of this economy." Lee Morrison, Reform MP from Saskatchewan, insisted that "climate is a cyclical phenomenon. It always had been and always will be." He accepted that the world is "slightly warmer" than 150 years ago but asserted that this warming proved nothing other than the existence of long-term climatic cycles. "For the life of me," argued Reform MP Dave Chatters, "I cannot understand how intelligent people can totally ignore any of the conflicting science on this issue." "Prudent action" was what he proposed, without specifying what that might mean.

Reform, with its political roots in prairie populism, its economic thinking driven by free enterprise capitalism, and its ideological opposition to government planning and regulation, sowed the seeds for the Canadian Alliance's antipathy to Kyoto and skepticism about climate change, attitudes that were inherited by Stephen Harper's Conservative Party. From an energy policy perspective, the Conservative Party simply absorbed the Canadian Alliance, since the Alliance provided the members from the energy-producing provinces (except Newfoundland). Harper himself had been fiercely against Kyoto and deeply skeptical of the science of climate change before becoming leader of the Canadian Alliance. As leader, he lit into Kyoto, urging repeatedly that the government not ratify the accord. "We will stand alone in the House, not just opposing ratification but urging blockage by the provinces and anyone else who is able, of implementing the accord and we will repeal the accord at the very first opportunity," he pledged in a Commons debate shortly before Chrétien announced ratification in 2002. In a newspaper article written for the *Globe and Mail*, Harper declared, "As an economic policy, the Kyoto accord is a disaster, and as environmental policy, it is a fraud."

KYOTO: HARPER'S "SOCIALIST SCHEME"

As part of the Alliance's campaign against Kyoto, Harper sent a fundraising letter to supporters that claimed that climate change was based on "tentative and contradictory scientific evidence. It focuses on carbon dioxide, which is essential to life, rather than upon pollution." Harper asserted that the oil and gas industry would be crippled by Kyoto, whereas Third World economies would be helped. "Kyoto," he warned, "is essentially a socialist scheme to suck money out of wealth-producing nations." With Chrétien's Liberals heading toward ratification, Harper pledged "to do everything we can to stop him . . . but he might get it passed with the help of the socialists in the NDP and the separatists in the Bloc Québécois."

Like Manning five years before, and with equal justice, Harper decried the lack of an "implementation plan" for Kyoto. The 2002 climate change plan was full of holes, he correctly argued, and it lacked any costing. Harper, like Manning, was right to ridicule Canada's Kyoto target, since by 2002 it was clear the country had no coherent plan to meet it. He estimated that 60,000 to 250,000 jobs would be lost, and expenditures and impacts on the economy would be between $5 billion and $25 billion.

Then he got onto the science, quoting the business community's favourite anti-Kyoto scientist, Richard Lindzen of the Massachusetts Institute of Technology. Experts, Harper asserted, had not identified the source of global warming, because the calculations involve "complicated and complex science that is far from settled." Kyoto involved only a small number of countries, left out giants such as India and China (to say nothing of the United States), imposed unfair burdens on a big, northern country such as Canada, and should be stopped. "Things must be done in a way that is consistent with the economic needs of ordinary people. That requires us to be consistent with the plans being developed by our provinces and our trading partners," Harper said. He meant that the federal government should move at the pace of Premier Ralph Klein's Alberta government and that of President George W. Bush in the United States – namely, a snail's pace, to perhaps slow, at best, the increase in GHG emissions, as opposed to reducing them.

In Kyoto's place, Harper suggested Canada "find CO_2 reduction targets consistent with long-term economic growth, encourage energy efficiency improvements, technological innovations and alternative energy sources to reduce the intensity of emissions, put our money into research and development, focus on actual pollution measures, and monitor the science to get a much better understanding of the extent of climate change and its causes." Having excoriated Kyoto and denounced Canada's target, Harper then proposed an approach so weak that by inference he heartened all those who did not treat climate change seriously and therefore preferred minimalist, cost-free measures.

BUSINESS AS USUAL

The Canadian Alliance's alternative became almost word for word the position of the Conservative Party in opposition and later in government. The position was essentially that of the business community: scale way back on emissions reduction requirements to balance "long-term economic growth," spend money on research and alternative energy, do nothing whatsoever about fossil fuel emissions except favouring intensity reductions, and wait patiently for science to develop a clearer picture of the problem and possible solutions. It was as close to a business-as-usual position on GHG emissions as could be devised without using the actual words. It was less, far less, than the Liberals were proposing, and of course Liberal policies, as we have seen, were failing completely even to slow the increase, let alone begin a decline, of GHG emissions. The Conservatives were further ahead than the Liberals, at least rhetorically, in placing GHG emissions in the context of smog and other threats to air pollution that were actually more threatening to ordinary Canadians than climate change; which was based on uncertain science and involved huge costs to taxpayers. Harper himself used to say that, as an asthma sufferer, he knew all about the problems associated with smog. Given his political judgment, it was not surprising that when the Conservatives drafted their campaign document, *Stand Up for Canada*, they spent more time promising to do something about air pollutants such as smog than they gave to GHGs. That party strategists paid more attention to smog reflected their belief that global warming bothered elites but not suburbanites and rural voters, the Conservatives' core political targets. A successful election campaign that featured only passing references to climate change seemed to vindicate their assumptions. That approach is what they had convinced themselves was better policy while in opposition, and that is what they believed for some months after forming the government.

But to their consternation and surprise, the issue of climate change, which they had belittled as a policy and political priority, suddenly stirred up a storm. In 2006, they found that the Canadian government could not, as Harper had promised, "tear up Kyoto" and wait for the science to become clear, for enough people were aware that the scientific evidence was by now irrefutable. They discovered, too, that the Canadian public would not allow the government to concentrate on other air pollution measures only, push CO_2 reduction targets off into the hazy distance, cancel Liberal programs without offering anything to replace them, scorn other countries that believed in Kyoto, and let the business community's long and successful campaign of denial and delay continue to frustrate serious emissions control measures.

A BAD CAREER MOVE FOR RONA AMBROSE

The appointment of Rona Ambrose from Edmonton as environment minister reassured Alberta that one of their own would protect the province from any onerous measures, the role Anne McLellan had played in the early Chrétien years. Ambrose had been elected only in 2004, after a brief career as a mid-level civil servant in the Alberta government. In opposition, Ambrose had been the federal-provincial relations critic, a title that suggested an importance the position did not have, since Harper made all the important strategic decisions about how the Conservatives would handle the provinces. Ambrose would be like many previous environment ministers, alone at the cabinet table with no natural allies in other departments: a second-tier minister for a secondary file.

Given the Conservatives' skepticism about climate change, and their conviction that the issue was of only secondary importance, Ambrose was essentially there to rag the policy puck. The trouble was, as bad luck would have it, that Canada was to assume the presidency of the Conference of the Parties, the United Nations body responsible for negotiating the next phase of GHG emissions under the Kyoto Protocol. Ambrose could try to run from the issue, or duck, but she could not hide – not when as minister she had to chair an international meeting in April 2006, in Bonn, just three months after the Conservatives' victory. Critics at home and observers in other countries were eager to know if "Canada's New Government," as the Conservatives insisted on calling themselves, would accept the Liberals' Kyoto target. Perhaps they would somehow jettison Canada's participation in Kyoto, or admit that the target could not be met, pay the potential penalties implied in the treaty for failure, but agree to make up after 2012 for what had not been done and accept additional GHG reduction burdens.

The Conservatives, pledged to a "made-in-Canada" policy they could not define and had not thought about in detail, showed at and before the Bonn conference that they lacked answers to these questions about Kyoto. Their confusion about what to do only deepened when their first briefings from senior civil servants told them how far Canada was from its Kyoto target, and how near to impossible it would be, short of spending billions to buy emissions credits overseas, to meet that target. "It is impossible, impossible for Canada to reach its Kyoto target," Ambrose said publicly after the briefings with senior officials at Environment and Natural Resources. But what to do?

At times, Ambrose said flaws existed in the Kyoto Protocol and Canada could work within it to make improvements. Having suggested a "made-in-Canada" solution as an alternative to Kyoto, Ambrose also suggested such a solution could be worked out within the accord. Then she said Canada would remain a party to Kyoto but would not respect Canada's engagements within it. After meeting with Bush administration officials and receiving an invitation from Australian prime minister John Howard during his official visit to Ottawa, Ambrose mused aloud about Canada joining the Asia-Pacific Partnership on Clean Development and Climate. This inspiringly named organization had been created by Kyoto refusenik countries (except Japan) to share information about climate change and sponsor research. The partnership, however, was toothless, without targets, penalties, or laws. The partnership was seen as a weak alternative to Kyoto, and the mere suggestion that Canada might join gave the impression that the Harper government would abandon Kyoto, or remain within the protocol but not abide by its stipulations.

At the Bonn meeting, whose purpose was to craft a post-2012 agreement, Ambrose quickly became the punching bag for environmental groups, pro-Kyoto countries, and opposition parties at home. Before she arrived, the government had sent an official statement to the conference, saying Canada would stay in the Kyoto process if the other signatories would give the country longer deadlines, voluntary targets, and exceptions for resource-based industries. And, by the way, added the Harper government, Canada would not be meeting its 2008–12 target. Taken together, the position made little sense. Canada would remain in a treaty process, the first phase of which it would not meet – and for the second phase it would demand conditions so lenient that no other Kyoto signatory would agree to them. The government believed in the Kyoto church but in none of its teachings or obligations.

Needless to say, a country and a minister presiding over a conference while uncertain about the conference's objective created a mixture of consternation in Bonn and confusion in Canada. Politically, however, the Conservatives were not worried about foreign consternation and policy confusion in the spring of 2006. Their internal polls continued to confirm their belief that climate change was no more or less important than other environmental concerns. Toxic materials, smog, water quality, mercury – these were just as important as climate change. Climate change remained an elite issue, and Rona Ambrose's job was to keep casting doubt on Kyoto while working on that "made-in-Canada" plan to be unveiled in due course. Meanwhile, to show critics that the Conservatives did care about the environment, Ambrose made announcements about getting mercury out of ignition switches for junked cars and announced with provincial counterparts a plan to increase to 5 per cent from 1 per cent the amount of ethanol in Canada's gasoline mix.

Unfortunately, these and other announcements neither cleared up the confusion over climate change nor hid the cuts the Conservatives were making to environmental programs, including ones dealing with climate change. Groups promoting the One-Tonne Challenge were told there would be no more funding. Then that program went entirely, along with the EnerGuide. Other Liberal climate change programs were axed, with none to replace them.

Money had to be found, after all, to fund the Conservatives' $250-million election promise of a tax break for transit passes. This foolish policy flew in the face of analyses from Environment bureaucrats who unavailingly warned their political bosses that the tax measure would have marginal impact. As a memo prepared in the deputy minister's office and leaked to the press told the government, "A wide range of data suggests that people are not very responsive to changes in transit fares. . . . While the ridership impacts of the tax incentives are not known with precision, analysis suggests they will be low." Taking money from targeted climate change programs and shifting it to such a marginally useful program as a tax break for transit riders made no policy sense whatsoever. But the tax break had been a political promise, revealed by Harper at a Vancouver photo opportunity on a bus. It was part of a series of campaign promises that involved giving money directly to voters. It could therefore not be changed, no matter how ineffective the policy.

Meanwhile, that promised "made-in-Canada" climate change policy was stuck in the bureaucratic swamps, and Rona Ambrose with it. The government had promised something in the fall of 2006, but plans just weren't coming together. The deputy minister of the environment, a man of ferocious drive but limited people skills, was shipped off to the World Bank. Ambrose's chief of staff and her communications adviser were also replaced, a sure sign in Ottawa of a floundering minister. Gradually, the Prime Minister's Office began taking control of the climate change file, a change of enormous significance.

Stephen Harper's government made Jean Chrétien's "friendly dictatorship" look like a lesson in delegation. In Stephen Harper's Sun King government, absolutely everything revolved around him, his message, his persona. Whereas at the start of the Conservatives' time in office, when climate change was considered a secondary file, Rona Ambrose was allowed to putter away on it, struggling to grasp its complexities, once deadlines loomed and public opinion began to get interested, the Prime Minister's Office took charge.

THE SUN KING TAKES OVER

That spelled the beginning of the end for Rona Ambrose as environment minister. But it was the end of the beginning for the Conservatives on climate change, because once the PMO took control, it signalled that for the first time the central driving agency of the federal government believed in the political necessity of doing something serious. No previous Prime Minister's Office – not Brian Mulroney's, Jean Chrétien's, or Paul Martin's – had ever run the file so tightly from the centre of government. And in the Canadian system, unless the prime minister either takes control or delegates to an experienced, trusted minister, the result is usually inertia. This particular prime ministerial interest did not come naturally to Stephen Harper, who, in his early months, still harboured the economist's doubts about the costs of serious action, preferred delay, and had listened to too many of his own speeches. His political soulmate, Australia's prime minister, John Howard, a winner of four elections, came to Ottawa and reinforced Harper's own instincts with advice about the flaws of Kyoto. Harper was focused on his new government's famous five priorities, among which climate change was certainly not one.

Sometimes, however, the great leviathan of public opinion bestirs in the depths for reasons difficult to fathom. So it happened with public opinion and climate change in 2006. An issue that had been near the margin of public concern for years steadily, then rapidly caught fire that year. By year's end, and into 2007, the broad subject of the "environment" had eclipsed the hardy perennial of health care as Canadians' number one concern. Or so pollsters discovered. Stéphane Dion did, too, for he captured the Liberal leadership by literally draping his supporters in light green on the morning of the convention vote, and he defined his candidacy primarily around sustainable development, especially climate change. He did not hesitate – he, whose plan as Paul Martin's environment minister had been so well intentioned but flawed and ineffectual – to call climate change the most important issue of our generation. First the Liberal delegates and then the Canadian public agreed.

The media, too, jumped on climate change, sensing a story of interest to readers and viewers. In the 1990s and early 2000s, a handful of excellent journalists such as Anne McIlroy and Alanna Mitchell of the *Globe and Mail*, Peter Calamai of the *Toronto Star*, and the producers and reporters at CBC's *Quirks and Quarks* had laboured to report, explain, and analyze climate change. Their efforts never stirred a public reaction beyond those already persuaded of the issue's importance. The *Globe and Mail*'s editorial board pooh-poohed climate change; the *National Post*, predictably, lampooned the whole issue, while its financial page editor and columnist Terence Corcoran led a weekly campaign against the very idea that climate change was real, let alone that anything should be done. (On one occasion, a cold, wet week at his summer cottage provided him with the conclusive proof that global warming does not exist.) Suddenly, in late 2006, editorial boards, columnists, reporters, and editors could not display their interest in the environment fast enough. Academics who had laboured in obscurity were suddenly in demand for analysis and commentary. Whereas the first report of the Intergovernmental Panel on Climate Change had drawn only a handful of reporters, the fourth report, in February 2007, attracted a standing-room-only crowd of reporters to a large room in central Ottawa.

The environment had experienced blips of public concern before – over the ozone layer or water quality, for example – but these blips never built any momentum to make the environment in general, or climate change in particular, a top-of-mind issue. Politicians used a rough rule of thumb: when the economy hummed, voters relaxed about jobs and could worry more about the environment, and vice versa. Environmental concerns were therefore cyclical, and politicians assumed that if a particular environmental issue arose, and they were seen to respond, the environment would settle back to its accustomed place among second-tier concerns.

The upsurge in public concern – at least, nominal public concern – caught the Conservatives flat-footed. What had been reckoned a secondary issue became *the* issue of the moment, and one for which they were politically and conceptually ill-prepared. They did issue, in the fall of 2006, a notice of intent to reduce "air emissions," which was to have been their climate change policy. It was panned, predictably, by the opposition parties in Parliament and environmentalists but by less politically engaged observers, too. The reaction quickly persuaded Harper that this first effort had fallen short, and he promised changes.

The first effort did fall short in important respects, but given how the Conservative Party had viewed climate change as recently as its coming to power less than a year before, the notice of intent was a startling document. Right away, the government scrubbed what had been the preferred Liberal strategies – and its own when in opposition. "Canada has historically relied on a variety of non-compulsory measures to reduce air emissions," it declared. "However, these have not proved sufficient to reduce the health and environmental risks across the country." That sentence, and the entire tone of the notice, seemed to indicate a determination to act on air pollution and GHGs. The automobile industry was put on notice that a new, stricter, mandatory regime would replace the voluntary regime for emissions reductions when it ended in 2010. There would be stricter regulations for a variety of engines and machines. The government also outlined important principles, including the achievement of "measurable reductions in air pollution that will produce health and environmental results." So far, so good.

Then came the core of the government's proposed approach, and it was sadly lacking. For GHGs, the government adopted the intensity-based approach for 2010–15, with a tougher approach for 2020–25, but still based on intensity targets. As critics quickly noted, this would still allow emissions to rise as energy output increased. The government set only one very long-term target: to reduce emissions by between 45 and 65 per cent from 2003 levels (not the original Kyoto target year of 1990) by 2050. Critics rightly pounced on this distant goal, demanding to know where were the interim, measurable targets and the policies that could effectively achieve them. The influence of the fossil fuel lobby could be seen not only in the intensity targets, but also in the proposal for a technology investment fund to which industry and possibly government could contribute.

Critics easily missed, however, the important – even unexpected – message from the Conservatives: voluntarism and subsidies were not enough. How and when stronger policies would be introduced would be negotiated over three years, but they were coming. When the Conservatives themselves acknowledged that their first effort had failed and volunteered to work with other political parties in Parliament to get a strong policy, it was a concession to their minority situation in the House, but also a recognition that the public wanted more and better.

Relaunching required a new minister. Rona Ambrose was identified with a policy now deemed inadequate. She had sealed her fate with several fumbling performances before House committees, in which she got facts wrong and had to be corrected by officials. She cancelled appearances and media interviews, and generally went to ground. The file had largely been removed from her anyway, as senior Environment Department and other officials now were working with the Prime Minister's Office. What the PMO needed was a salesperson, and Ambrose had become such a political liability that the media began a death watch on her career at Environment.

When the end came, and she was switched into federal-provincial relations (another file completely run by Harper), the obituaries of her performance at Environment were cruel, but only somewhat deserved. She had been given few directions, had been surprised like the rest of the government by a public mood shift, and had been handed little guidance from a party platform that had relegated climate change to the margin. She had had to deal with a prime minister who did not consider the issue of real importance, and had himself been a climate change skeptic.

PART II

GETTING OUT OF THE MESS

Options and Solutions

CHAPTER FOUR

A Tour of the Options

The challenge of climate change is daunting; the understanding required to get policies right in Canada so that emissions go down is considerable. Canada's population and economy are growing, consuming more energy. This is a basic fact, if uncomfortable for some: Canada is going to use more energy, not less. Our climate change problem is an emissions problem, rather than an energy one. If policy focuses on exhorting or otherwise influencing Canadians to use less energy, it will largely fail as an answer to climate change even though reducing energy use is desirable; if it focuses on lowering emissions through regulatory or market-based policies, it can work. This country is a major energy producer, but the kinds of energy that provinces produce range from low-emissions to high-emissions.

Governments must try to navigate between environmental and business lobbies, to say nothing of developing policies that the public will accept. Then there are international considerations, since global warming is a worldwide challenge. Even if Canada had not signed and ratified the Kyoto Protocol, Canadians would be obliged to consider what other countries are doing, and what they intend for the next round of Kyoto after 2012. As it is, Canada did ratify Kyoto and therefore undertook obligations to reduce GHG emissions by 2008–12.

Governments – and the public – often mix up policy and action. Governments like promising to do things: to "generate more energy," to "enhance energy efficiency." These are actions, whereas what we should expect from governments are policies that will produce actions by others. Governments can actually do very little, beyond changing their own operations. Rather, it's on the effectiveness of their policy options and choices that they should be judged. Rhetoric is easy and cheap, and Canada has certainly had plenty of rhetoric on the subject of climate change. What's needed instead now is some cool thinking about hot air, starting with the four major categories of action that government is trying to promote through better policies: land use, energy conservation, energy substitution, and GHG capture.

LAND USE: A PROMISING AREA

The first option requires fresh thinking and action about land use, especially management of soils, their vegetative cover, and even animal husbandry practices. Non-energy-related GHG emissions from agriculture and forestry are estimated to be about 10 per cent of Canada's total emissions, with agriculture alone accounting for about 9 per cent.

Different land uses absorb or emit different amounts of carbon, the bad stuff that would otherwise be in the atmosphere. Agricultural land converted to forest is more likely to store or absorb more carbon. Some forests sequester more carbon than others. Some cropping and soil management practices do better at sequestering carbon. Studies have shown that reconverting significant sections of the northern Canadian prairies back to their forested state would store a lot of carbon. The trouble is that while some of this land is currently of marginal quality because of temperature constraints, its agricultural value could increase significantly if a warming climate improves growing conditions for crops.

Animal husbandry – the raising of cows, pigs, sheep, all of them – generates nitrous oxide as well as methane. Such emissions might be reduced with different feeding and grazing practices and by control of manure to reduce methane release. When decomposing in an enclosed pile, manure emits methane. One possibility is to capture the methane and use it for energy, as they do in China, India, and other developing countries. Burning this form of biomass lowers methane emissions and produces energy to reduce the combustion of fossil fuels – a double benefit. Low-till farming keeps more carbon in the soil, and organic farming uses less commercially manufactured fertilizer.

Another kind of land use – urban and industrial landfills – also produces GHGs that can be curbed. Biomass and some other materials emit methane as they decompose in these landfills. As with manure in agriculture, these methane emissions can be captured and combusted to produce electricity and heat, as is done in some Scandinavian countries. Some municipalities or independent electricity companies are already starting to pursue this option in Canada, which is profitable under certain conditions.

ENERGY EFFICIENCY AND ENERGY CONSERVATION

The second option requires decreasing energy use, just about everybody's politically favourite option, especially among environmentalists. But businesses like this option, too, because it incorporates their demand for recognition of their "intensity" improvements while focusing attention on what consumers of energy must do. Decreasing energy use can be accomplished through "energy efficiency" – using more efficient technologies in buildings, factories, vehicles, appliances – and "energy conservation," which means changing behaviour so that these devices are used less frequently. Purchasing a more efficient light bulb (and remembering to install it!) represents an improvement in energy efficiency; turning the lights off when not in a room represents energy conservation. In a home, a standard incandescent light bulb uses 60 watts of electricity to produce 740 lumens of light energy, whereas a compact fluorescent light bulb might use about 15 watts to produce the same 740 lumens. One device is therefore four times more energy-efficient than the other. Buying a car that uses less fuel would be energy-efficient; driving it less would be energy conservation.

When people speak about an improvement in energy use, they usually mean both changes in conservation behaviour and advances in technology efficiency. A really determined effort to decrease energy use would lead to a multiplicity of changes including, to name a few, more efficient buildings and devices, expanded public transportation or more use of existing systems, co-generation of electricity and heat in industrial plants and buildings, and urban planning that increases density to reduce the distance travelled for work, shopping, or leisure.

In industry, firms have become generally more efficient at using energy to keep costs down. They can buy more energy-efficient machines – electric motors, boilers, conveyers, or whatever – or they can get greater efficiency when converting energy.

For example, co-generation involves a combustion system that produces both electricity *and* heat that can be useful for industrial or other purposes, like heating commercial and residential buildings. The high-quality heat from burning fossil fuels or biomass (wood or straw) in a boiler produces steam that spins an electricity-generating steam turbine. Although much of the steam energy is lost – it is converted into electricity – there is still some heat from steam exiting the turbine. In industry, this can provide a thermal energy service, like cooking wood chips into pulp in a pulp mill. This combined production of electricity and heat can achieve an energy conversion efficiency of 90 per cent – in other words, 90 per cent of the energy used (the fuel burned in the boiler) is converted to energy output (the electricity and heat). This kind of co-generation is often called "combined heat and power." There are some places in Canada, but many more in northern Europe, where town and city developments have been deliberately designed so that electricity can be generated locally while the lower-temperature waste heat is used to provide space heating or water heating in surrounding buildings.

Energy conservation comes in many forms. We could set the thermostat to a slightly lower temperature in winter during the day, or when we sleep. We could turn off lights in unused rooms. We could wash our clothes in cold water. We could cycle or walk more, and use the car less. These reductions in energy use do not involve buying more efficient devices; they mean using existing devices less, or changing behaviour patterns, without diminishing our quality of life. These are the kind of changes various Canadian programs of public information and exhortation have tried to impress upon Canadians, with minimal results thus far.

Obviously, the potential for adopting a conserver lifestyle is enhanced if urban areas are developed wisely, with conservation in mind. If streets are people-and-bicycle-friendly, if commercial and retail establishments are within walking and cycling distance, the chances rise that society's overall energy use will decline through energy conservation in addition to acquiring more efficient devices.

Energy efficiency improvements could also come from reduced flows of goods and services. Using less fertilizer in agriculture reduces its industrial production and therefore the associated energy use. Consumers who purchase fewer or lighter material objects reduce industrial energy requirements for the production and transport of goods. More recycling helps, too, because recycled materials require less energy to convert into usable products than do new materials.

Energy efficiency advocates can be found along a spectrum. Some advocates are technically minded, focusing on designing and distributing more efficient products and services; others reject in whole or in part our material-intensive lifestyle and yearn for a return to the lower energy demands of an earlier, simpler era. Regardless of where they line up, most people intuitively believe enhanced energy efficiency to be a highly desirable way of reducing GHG emissions. Energy efficiency seems like the obvious choice – the least painful choice, anyway – for curbing GHG emissions. The greater the energy efficiency, the fewer difficult debates we will have about nuclear power and where to dam rivers for hydroelectricity, the less vulnerable we are to oil and natural gas prices, the lighter the impacts on land and water from coal and uranium mining, oil and gas development, and oil sands production. And, at first glance, energy efficiency potential appears enormous. Some people suggest that society could easily, even profitably, reduce energy use by 25 to 50 per cent. Amory Lovins, perhaps the best-known energy efficiency advocate, sometimes asserts that energy use can be lowered by 75 per cent – profitably!

Looks, alas, can be deceiving. Increasing energy efficiency enough to lead to falling energy use can be difficult. We have been earnestly trying to reduce energy use for more than two decades with little success. Governments, for example, have offered financial incentives to buyers of hybrid cars, and consumers have been told that they will do the environment a favour by selecting models with lower greenhouse gas emissions. The actual evidence is so counterintuitive, however, that many energy efficiency advocates, along with politicians, journalists, and members of the general public, cannot seem to accept it. They often do not want to hear it. This reluctance is one of the biggest challenges to the design and implementation of effective policies to reduce GHG emissions. This same reluctance has also played a role in explaining the failure of policy efforts in Canada, a point we have already described and will explore further later.

SWITCHING FUELS AND PROCESSES

The third option – after land use changes and energy efficiency improvements – is "fuel switching," which encourages households and businesses to switch away from fuels and technologies that cause GHGs. Renewable forms of energy, in contrast to fossil fuels, are seen as GHG-free, including solar power, generating electricity from photovoltaic arrays, and heating domestic hot water with rooftop panels. Wind power produces electricity from massive turbines hooked to grids, an upgrade on the old wind-driven waterpumps that once dotted the rural landscape. Hydro power is found in abundance throughout most of Canada. Geothermal energy can be useful, where it is available, as in Iceland or New Zealand. Oceans are receiving more research attention, as scientists wrestle with getting energy from tidal movements, waves, and currents, or from heat gradients between the surface and the deep ocean.

The renewable stable also includes biomass, such as wood waste from forestry or residual crop waste from agriculture. This form of renewable energy is not considered a net emitter of carbon dioxide when combusted raw or after being converted to liquid fuels such as biodiesel or ethanol. This is because the carbon released from biomass when it is used is presumed to be re-extracted from the atmosphere by future biomass growth, in the form of more trees or grain crops, a virtuous closed loop with useful energy for humans but no net increase in atmospheric carbon.

Finally, and controversially, there is nuclear energy, essentially free of GHG emissions but never free from political controversy, in North America and Europe especially.

Fuel switching does not just mean moving away from fossil fuels; it can sometimes mean switching among them. Refined petroleum products such as gasoline, diesel, jet fuel, propane, butane, bunker fuel, and heating oil emit less CO_2 than coal. Natural gas emits even less. Switching an electricity plant from coal to natural gas would decrease GHG emissions. One of the ironies of modern times is how Britain made a dramatic reduction in GHG emissions under Prime Minister Margaret Thatcher, not a renowned environmentalist. In order to shatter the strength of the coal miners' unions, she allowed privatized utilities and their subsidiaries to build new electricity generation plants fuelled by North Sea natural gas. Any such switch anywhere would make excellent sense from a climate change perspective, but since natural gas prices have risen more than coal prices in recent years, price changes can unfortunately reverse the benefits of fuel switching.

Companies can also switch industrial processes to reduce GHG emissions without changing fuels, although sometimes this switch can be costly. For example, the conventional procedures involved in making cement, aluminum, and chemicals emit substantial GHGs, but different processes can, in some cases, save energy and produce lower emissions. Indeed, this is happening in the aluminum and chemical industries. GHGs also result from leaks of methane from gas lines, and from venting and flaring of gases in oil and gas extraction. These emissions can often be reduced with slight changes in practice, detecting and preventing leaks, and eliminating flaring. No fuel switching results from these changes, just better procedures that restrict emissions. And the oil and gas industry, to its credit, has been reducing flaring.

Not all countries have the same opportunities for fuel switching. Most countries use refined petroleum products for fuelling the transportation sector, so they are all in roughly the same situation, although many, including Canada, are racing to increase the use of biofuels in vehicles. When it comes to electricity generation, however, some countries (such as France) depend on nuclear power, others on coal, natural gas, or some combination of these. Canada, because of its geography, already generates more than 60 per cent of its electricity from hydro. The federal government's post-war efforts to develop and market its own nuclear technology (the CANDU reactor) led to some nuclear power generation in Canada. As a result, only 27 per cent of Canadian electrical power is generated from fossil fuels, especially coal. Other countries that are far more reliant on fossil fuels of one kind or another than Canada have greater possibilities for fuel switching. If countries are asked to reduce emissions by similar percentages, it could be more costly for the Canadian economy to achieve the same percentage reduction; it must soon pursue emissions reductions in transportation, which are comparatively more costly than reductions in the electricity generation sector.

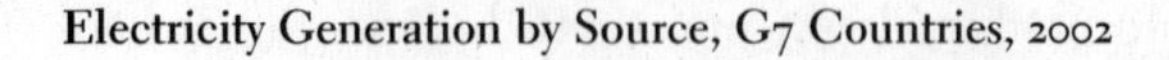
Electricity Generation by Source, G7 Countries, 2002

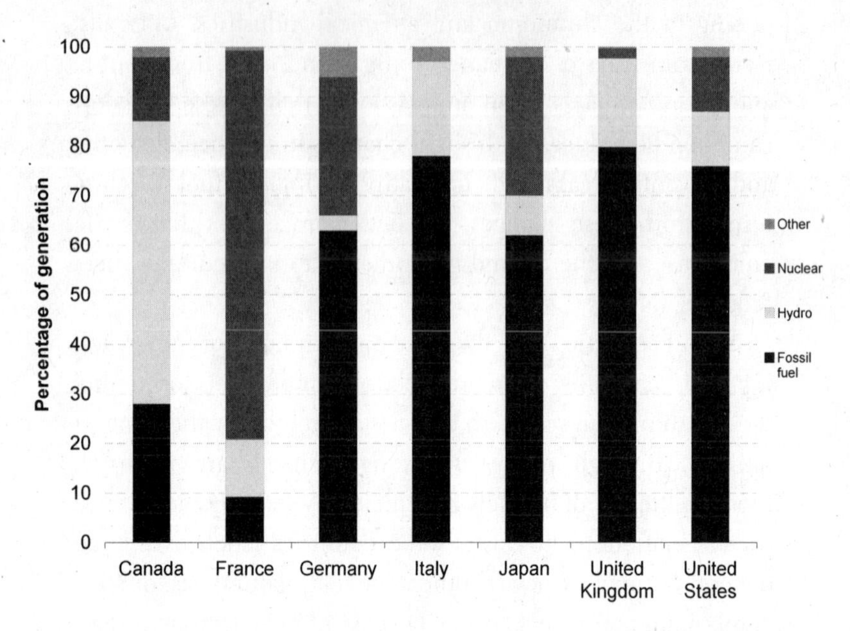

Source: United States Department of Energy.

CAPTURE AND STORAGE

Finally, the fourth option for reducing GHGs is emissions capture and storage, an approach akin to past efforts to use fossil fuels more cleanly. Emissions reduction and capture technologies have been developed for almost all fossil fuel combustion, including different burner technologies to reduce emissions of soot and nitrous oxides, which contribute to air pollution and acid rain. Catalytic converters are now common to virtually all internal combustion engines. Major industries such as coal-fired electricity generating plants apply smokestack scrubbers and chemical processes to capture particulates and sulphur dioxides that would otherwise contribute to acid rain.

The shift in scientific and public opinion against climate-unfriendly CO_2 emissions caught the fossil fuel industry and other industries by surprise. Many responded with denial; others fought for delay in taking action. Recently, companies have begun to study technologies – some already in use – that capture CO_2 emissions and store them safely away from the atmosphere in underground sedimentary layers. This promising avenue for preventing emissions does not represent a huge conceptual leap from other efforts to prevent pollution from fossil fuels, although business is eager to offload as many of the development costs as possible on government.

The option of capture and storage of GHGs bothers some environmentalists. They oppose this approach, or at least play down its importance, because these technologies accept, rather than challenge, the high-emissions character of our energy and economic systems. Environmental purists argue that a sustainable economy depends on humanity reorganizing itself so that no pollution is generated, and that any by-products created by human activity are completely harmless to the environment; as an example, they point to biodegradable wastes being easily assimilated by natural systems. They prefer energy efficiency and fuel switching, from fossil fuels to renewables. They press governments to encourage these changes, while being less receptive to end-of-pipe responses that contemplate the continued use of fossil fuels.

As for the hydrogen dream, we should pay no attention when advocates insist that the hydrogen age will replace the fossil fuel age. Some writers tout hydrogen as the world's long-term answer to climate change, but these books cause confusion. Hydrogen, like electricity, is a "secondary energy" – that is, produced from one of the three primary sources: fossil fuels, nuclear power, and renewables. Portraying hydrogen as a saviour is like describing electricity as a saviour. They are both zero-emissions secondary sources of energy, but something has to be used to make them. We could use fossil fuels, or nuclear power, or renewable energy. Will hydrogen become a partner with electricity in a clean future? No one is sure, so great are the uncertainties surrounding hydrogen. We should be happy either way. If we are going to reduce our GHG emissions dramatically, the most likely long-term scenario is that most of our end-use devices will run on a variety of electricity, hydrogen, biofuels (like ethanol, biodiesel, and biogases), and heat (like steam and hot water distributed to buildings for space heating).

The technological and economic prospects for carbon capture and storage in the fossil fuel supply industry, including electricity generation and enhanced oil recovery, are quite good. Fossil fuels now provide about 85 per cent of global energy needs. Coal is the most plentiful and the dirtiest fossil fuel but will be extensively used around the world for decades to come. About 40 per cent of global electricity production comes from coal-fired plants. These plants account for about 30 per cent of energy-related GHG emissions. Gradually converting the world's coal-fired electricity generation to carbon capture-and-storage technologies is by far the most promising avenue for GHG emissions prevention.

One option would be to continue burning coal but to extract CO_2 from the power plant's smokestack. This "scrubbing technique" can be integrated into new coal-fired power plants and perhaps eventually retrofitted into existing plants. A variation on this option involves introducing pure oxygen into the combustion chamber, increasing the concentration of CO_2 in the smokestack and thus reducing the costs of extracting it, but requiring more energy to produce the oxygen.

A different technology would convert coal into a synthetic gas, which is separated into a hydrogen-rich fuel to generate electricity and a stream of CO_2. Other undesired by-products (particulates, nitrogen oxides) are ruled out by virtue of the process or captured at some point, too (sulphur dioxide, which contributes to acid rain, and mercury, which can be toxic in certain chemical forms and concentrations). Variants of this gasification process have been used for some time in oil refineries and fertilizer plants, and in the conversion of coal to synthetic gasoline and diesels, which was developed commercially in South Africa under the boycott caused by the regime's apartheid policy and is still done today.

Oil and natural gas could also be used in this way. They have higher hydrogen content and great energy density, so they can be converted into hydrogen and electricity more efficiently than coal – but they are more expensive fuels. Coal, being low-cost, is so plentiful and so widely distributed around the world that it has received the most attention.

What is especially interesting about zero-emissions coal-fired generation of electricity is that the technologies have been used individually in other commercial applications for decades, but they have not been put together before. That explains why the U.S. government has set up its "FutureGen" initiative, a $1-billion coal gasification plant that would generate 275 megawatts of electricity and also serve as a laboratory for producing hydrogen from coal and for CO_2 capture and storage. Several other governments have launched initiatives to build demonstration plants; some major electricity companies have announced plans to build larger, commercial-scale plants.

Zero-emissions use of fossil fuels depends on GHGs being permanently and safely prevented from reaching the atmosphere. Here, too, we have a great deal of experience, because some industries have been required to store various solid and gaseous wastes safely or convert them into marketable products.

Carbon can be stored in various ways: as a solid on the surface of the earth, as a liquid deep in the floor of the ocean, or in geological formations as gas, liquid, or solid forms after a chemical reaction. Storage in these geological formations is the most likely option, since we already have technological, commercial, and regulatory experience with this approach. For several decades, the oil and gas industry has transported CO_2 and injected it into underground geological structures. In more than 70 sites worldwide, CO_2 is injected into oil reservoirs to increase pressure to enhance oil recovery. CO_2 injection is also a means for enhanced natural gas recovery and for dislodging methane from deep coal deposits as part of coal-bed methane production. Finally, CO_2 can be injected into saline aquifers deep in the earth. In Canada, CO_2 is already disposed of in this way along with hydrogen sulphide in order to produce marketable natural gas; there are more than 50 sites using this acid gas injection technique in Canada today. Thus, all three forms of fossil fuels can be converted to clean energy, with the CO_2 by-product stored deep underground.

Conversion of Fossil Fuels into Clean Energy with Carbon Capture and Storage

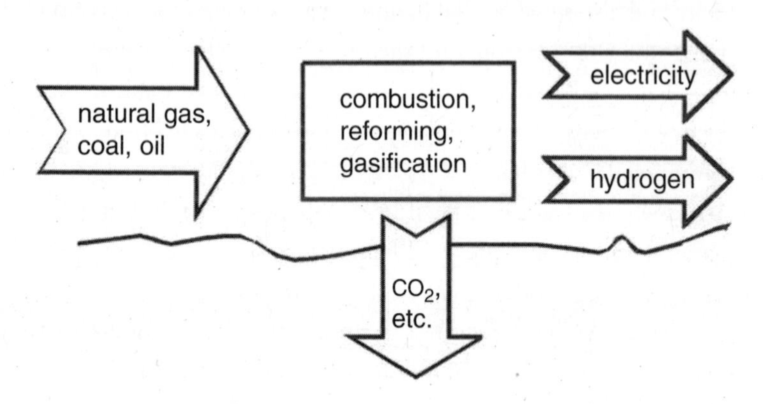

A commercial demonstration of world-scale significance for the capture-and-storage option already exists in Canada. Since 2000, a plant in North Dakota has been shipping CO_2 by pipeline to Weyburn, Saskatchewan. (There are more than 3,000 kilometres of pipelines safely carrying CO_2 from various sources around North America to enhanced oil recovery projects.) There, the CO_2 is injected into an aging oil field whose yield has been increased in this manner by 30 per cent. The North Dakota plant happens to be a coal gasification facility that produces a hydrogen-rich gas for industrial uses and a stream of CO_2 as a by-product. Instead of being released into the atmosphere, 20 million tonnes of CO_2 is being shipped over the next 30 years to Weyburn. Industry, governments, and researchers are closely monitoring the project because it integrates all of the essential components of a zero-emissions fossil fuel system: coal gasification, production of a hydrogen-rich fuel, capture of pure CO_2, a long CO_2 pipeline, and geological storage of CO_2.

This project and other economically attractive ones indicate the feasibility of a concerted effort to sequester CO_2 in depleted oil and gas reservoirs. Promising as this form of carbon storage is, current and future depleted reservoirs have a combined carbon capacity of only 300 billion to 600 billion tonnes of carbon, which may seem like a lot but is not nearly enough to contain all carbon from fossil fuels if oil, coal, and natural gas continue to dominate global energy through this century and beyond.

Researchers and some practical experiences suggest other possibilities. Much more plentiful and suitable storage sites are to be found in deep saline aquifers, which lie under certain rock formations at depths greater than 800 metres – deeper than the typical freshwater aquifers found at 300 metres or less. Saline aquifers, contrary to a common misperception of the word *aquifer*, are not underground pools of water. Rather, they consist of porous rock infiltrated with highly saline water. Depending on pressure, porosity, and other conditions, these sponge-like deep saline aquifers could absorb large quantities of CO_2, which would be condensed into a liquid-like form.

Efforts to estimate the total CO_2 storage capacity of deep saline aquifers are still crude, but the capacity is known to be large. Initial estimates of such aquifers – two-thirds are onshore, one-third offshore – suggest the possibility of their holding all of the planet's estimated carbon endowment in fossil fuels. Western Canada has many of these deep aquifers. Ontario's geology is less promising, but there is some capacity under the floor of Lake Erie that is now being studied by businesses and academics.

The petroleum industry knows about saline aquifers and the dynamic properties of injected CO_2. But until concerns developed recently about climate change, little interest was shown in CO_2 sequestration in saline aquifers – for the simple reason that nobody cared much, if at all, about GHGs. Then Norway shook things up with a carbon tax of $30 to $35 per tonne in the early 1990s ($75 in some sectors), and this motivated the Sleipner project in 1996. This project involved injection of CO_2 into a deep saline aquifer below the North Sea, not for enhanced oil or gas recovery but to avoid the carbon tax. (By contrast, the Liberal government in Canada, under pressure from the oil and gas industry, had promised that if a price were placed on carbon, it would be only $15 per tonne.)

At the Sleipner project, the CO_2 source is a reservoir of natural gas about 300 metres below the sea floor whose high CO_2 content must be reduced to meet market specifications. A chemical solvent is used to separate CO_2 from the natural gas and then inject it into a saline aquifer 1,000 metres below the sea floor. The solvent is continuously recycled in the process, and the cleaned natural gas is shipped by pipeline along the sea floor to northern

Europe. It works well. Several new projects are in the planning stages or under development in Norway, Algeria, Australia, Britain, and the United States.

The scale of the challenge and of the possibilities for the United States – and, by extension, for other countries with abundant coal reserves such as Canada – was underlined by an interdisciplinary study by professors at the Massachusetts Institute of Technology in early 2007. They noted that the 50 per cent of America's electricity that is generated by coal-fired plants produces 1.5 billion tonnes of CO_2, whereas the Sleipner project, the largest sequestration project in the world, buries 1 million tonnes a year. Coal, being plentiful and cheap, will continue to be used massively, so that "carbon capture and storage is the critical enabling technology that would reduce CO_2 emissions significantly while also allowing coal to meet the world's pressing energy needs." Governments have to work urgently with industry to finance demonstration projects beyond FutureGen, and not just in the United States, because geological formations vary widely. The MIT team took note of the Weyburn project in Canada as demonstrating that large-scale CO_2 injection could operate safely, but it said that many more demonstration projects were needed to work through the many technical and regulatory issues surrounding carbon capture and storage. The sooner the United States and other industrialized countries master these issues, the faster the technology can be adopted by developing countries such as India and China. Carbon capture and storage is critical for marrying the need for energy from coal with the need to combat global warming.

As always, there are risks involved in carbon storage, but preliminary work suggests that these are unlikely to be deal-breakers for the public. There are costs, too, and companies have been wary of incurring them, as long as public policy did not require them to act. They preferred voluntarism and moving at their own speed. And environmental purists often criticize carbon storage because it accepts that fossil fuels will be a mainstay of economies for many decades to come, whereas they want a much faster move away from these fuels than most Canadians. The resistance, or at least lack of enthusiasm, from environmentalists and business, coupled with government timidity, has left Canada behind in exploring carbon capture and storage possibilities that are fruitful solutions to the GHG challenge.

We should not close our eyes to this potential, because if we are to reduce GHGs to the extent that scientists say we must, we will need to move from combustion of gasoline and diesel for cars or natural gas for homes to using only electricity and hydrogen, some of which might be produced from coal, oil, or natural gas in conversion processes that capture and store all of the carbon in these source fuels. In theory, we could think of technologies that capture CO_2 wherever it is produced, whether at an oil sands plant or in a personal car or a home furnace. In practice, the costs of capturing, collecting, and disposing of CO_2 from smaller end-use devices are expected to be very large – at least until a technological breakthrough occurs.

Industry is therefore focusing initially on carbon capture and storage at large facilities. Only a few experts were thinking and talking about carbon capture and storage a decade ago; now, companies are working on the technology, politicians are floating the idea, and bureaucrats are studying the technological and legal issues. Carbon capture and storage may well be a major part of any national attack on GHG emissions.

SNAKE OIL AND SHORT-TERM SOLUTIONS

Canadian governments, in making international commitments about global warming and climate change, considered all four types of actions when drafting domestic policies – land use changes, energy efficiency, fuel switching, and controlling emissions. Conceptually, governments understood the options, but when it came to adopting policies that would produce useful actions by companies, other governments, or individuals, the policies were ineffective.

Right now, GHG emissions do not carry a price. Unlike many European governments, the Canadian government imposes no fee in the form, say, of a tax on emissions. The atmosphere is therefore considered a free trash bin into which we can dump GHGs. The result has been the classic tragedy of the commons, when everyone contributes to common degradation, but since no one action has been decisive in creating the problem, no one feels responsible for past or future action. Nothing is therefore done. In the case of climate change, no direct benefit accrues to individuals or companies from reduced emissions. So without strong government policy, such as putting a price on emissions or restraining them by regulation – policies that mandate action – very little will happen. This has been the Canadian climate change story thus far.

Action involves changes, sometimes dramatic, to infrastructure, buildings, industrial plant, equipment, and land use practices. These capital stocks develop over many years, sometimes many decades. They do not turn over quickly. Cars in Canada last about 13 years; electric power plants last about 50 years; buildings even longer. It would be enormously expensive, and usually impossible, for governments to dictate that equipment be retired before the end of its natural life. Cities have grown over decades into their current configuration, with their downtowns surrounded by massive urban sprawl. Their fundamental shape is set, and will be changed only incrementally, for better or worse, with each planning decision.

Significantly reducing GHGs under these circumstances will demand decades of effort. These days, plenty of policy snake oil salesmen are around peddling short-term fixes. Everybody seems to have a solution to offer. Meanwhile, Canadians, like the rest of humanity, continue to use more energy because they continue to value the comfort, entertainment, mobility, and status that greater energy use can bring. Environmental purists don't like to admit – indeed, they fight against – the observable fact that Canadians, like other humans, continue to extract, process, and consume more fossil fuels because these fuels can bring high-quality energy at the lowest-cost option, and we will do so for a very long time. The problem, given this certainty, is that as long as GHG emissions are free, humans will keep dumping them in the trash bin of the environment.

How to stop or substantially reduce that dumping is the core of the challenge. Thus far, Canadian governments have mostly talked about "targets," "time frames," and "action plans." Governments have excelled at setting targets but failed at getting results.

THE FOUR CRITERIA

Now that climate change has become central to public discussion in Canada, citizens should evaluate what policies are on offer by using four criteria: effectiveness at achieving environmental targets, economic efficiency, administrative feasibility, and political acceptability.

Ideally, all four criteria should be considered together. A policy might be politically acceptable but relatively ineffective in achieving an environmental target, as most of those so far introduced in Canada have been. Or a policy might be effective, as a cap-and-trade emissions system would be, but administratively unfeasible, if poorly designed. Or a policy might be economically efficient and environmentally effective, like a GHG tax, but politically unacceptable depending on the design. Moreover, Canada's federal system complicates both the development and the implementation of any policy. But Canadian policies are helped by moving in concert with those of other countries, because international pressures or obligations spur domestic action. Finally, GHG reduction policy requires making decisions in an area filled with uncertainties. If perfect certainty is any policy's first requirement, we can forget about accomplishing anything.

A SPECTRUM OF POLICY OPTIONS

With the four evaluation criteria in mind, and remembering the Canada-specific constraints on action, think about the broad options available to governments. There are actually a limited number of broad options, although many variations within each. Think of a spectrum of options: compulsory at one end, voluntary at the other.

"Command-and-control regulations" are technology or performance standards enforced through stringent financial or legal penalties. This approach, the most compulsory of the options, dominated environmental policy in the 1970s. Economists criticize command-and-control regulations because they impose uniform solutions on different situations: for instance, requiring identical equipment for all emitters, although the costs of emissions reduction might differ for each firm or individual. This approach also penalizes, or at least does not encourage, companies that want to go beyond the regulations. But regulations can eliminate bad choices, averting the high cost of learning about poor technologies the hard way. A company can hire consultants to identify options, and presumably the consultants will weed out the poor technology. A consumer whose fridge or furnace breaks does not have this luxury or information. Regulations banning poor technology can be a consumer's best friend.

"Market-oriented regulations" set an aggregate regulatory requirement on an entire economy or sector. This approach allows choice. Companies can choose whether to act according to the regulation or pay others to act on their behalf. This choice is exercised through the freedom to exceed or fall below a GHG reduction standard. Exceed the standard and be in a position to sell the excess; fall below and be forced to buy an excess. This kind of regulation greatly improves upon the command-and-control approach because it allows for flexibility and encourages economic efficiency. If the objective is to reduce GHG emissions over an entire sector or a large geographic territory, market-oriented regulations can work well. Something mandatory for individual firms is required, however, if the environmental problem is limited to one place or caused by one firm.

Market-oriented regulations for GHGs can be of two kinds. Under a system of emissions caps with tradable permits, the government allocates permits so that the total of all permits equals the government's emissions targets – the cap. Government targets have meaning under this system – as long as the penalties for not holding permits are strong enough to ensure compliance. Permit-holders are free to choose between reducing emissions to comply with their permits, and purchasing surplus permits from those who have exceeded their past required reductions. Governments can allocate the permits initially either by auction or by free distribution to emitters based on their past record of emissions. This latter approach compensates a region for its historical dependency on fossil fuel exploitation. If a region such as Alberta were granted permits in line with its historical emissions, and if it turned out that costs of reduction were cheaper there, Alberta could end up selling permits to other parts of Canada so that its share of Canadian costs would be comparable to those of other regions.

Another form of market-oriented regulation is "obligation with tradable certificates." Under this system, the government sets obligations for manufacturers or retailers. The government wants certain results, such as sales of low-emissions technologies or non-carbon forms of energy. Each desired kind of sale is awarded a certificate. These certificates must then equal the manufacturer's or retailer's obligations in order to avoid government-imposed penalties. Sales above obligations earn excess certificates. These can then be sold to those who have not fulfilled their obligations. For example, a renewable portfolio standard is used for the electricity industry in about half the U.S. states and many European countries. Sellers must ensure that a minimum share of the electricity they market is from renewables. If they fall short, sellers pay a penalty or buy renewable energy certificates from sellers with excess.

California's vehicle emissions standard is also a system of obligation with tradable certificates. This widely heralded (and criticized, at least by auto manufacturers) policy requires automobile sellers to ensure that a minimum share of their vehicles are low-emissions or zero-emissions. If not, sellers must make up for falling short by purchasing certificates from others who have exceeded their obligations.

These regulatory policies have not been favoured in Canada, even though they are essential for improving Canada's emissions record. Similarly, Canadian governments have shied away from taxes on emissions as too politically controversial. Polls still show a majority of Canadians opposed to taxes on such items as gasoline, and politicians of all stripes know it. On the spectrum of policies running from compulsory to voluntary, taxes are close to market-oriented regulations. They are compulsory in the sense that the emitter must pay a tax, or pay for other actions that reduce emissions, thereby avoiding the tax. They are non-compulsory in the sense that the emitters have flexibility to decide whether to pay the tax or invest money to reduce their GHG emissions and pay less tax.

Several European countries use GHG taxes. Unlike an emissions cap with tradable permits, an emissions tax does not guarantee a particular level of emissions, because no one knows how many, or which, emitters will change their behaviour to avoid or minimize the tax. A tax provides certainty to business and consumers about costs, but not about environmental effectiveness. GHG tax revenues can be used in various ways, such as reducing other taxes so that the net effect is revenue-neutral, or investing in clean technologies, or any other purpose deemed fit and politically saleable by the government.

Then there are subsidies, the Canadian standby in climate change policy and many other fields of economic activity. Subsidies come in all forms: rebates, grants, low-interest loans, so-called tax exemptions. Subsidies in theory are designed to improve financial returns and thereby provide incentives to businesses and consumers to reduce emissions. This approach always appears non-compulsory, but the funds have to come from compulsory sources of taxation. That is what subsidies are all about. They take revenues from compulsory revenue sources and spread them via government discretion for social, economic, regional, or political purposes.

The fifth policy option available to government, and politically the easiest, is disseminating information. Governments can encourage, exhort, or shame firms and households to reduce GHG emissions. Information campaigns can appeal to self-interest, morality, global obligations, and even intergenerational fairness. They can be of widespread applicability or be targeted through energy efficiency or emissions labelling. If you go shopping for a fridge or stove these days, labels will help tell you which product is most energy-efficient. On a government website, you can learn about all the things you can do to reduce your own output of GHG emissions.

NON-COMPULSORY POLICIES DON'T WORK

Thus far, Canadian governments have preferred the last two options: subsidies and information, the options at the non-compulsory or voluntary end of the spectrum of choices. They have their place, but they are addendums or supplements to what Europeans have done. They are useful only if you combine them with compulsory policies such as command-and-control regulations, GHG taxes, and market-oriented regulations. Effective action against climate change must deploy these policies to be effective.

Canada's almost total reliance on non-compulsory policies is the main reason why the country has failed to limit the increase of emissions, let alone start reducing them. The ineffectiveness of subsidies and information alone, however, is not just the Canadian experience. A growing number of studies conducted by some of the world's most respected energy experts draw the same conclusions about the experience elsewhere. Paul Joskow of MIT showed that efficiency subsidy programs by electric utilities were not nearly as effective as utilities and efficiency advocates claimed. Robert Stavins of Harvard has found evidence of similar problems, as has Kenneth Train of Berkeley. What lessons can be learned from these experiences?

FREE RIDERS

One lesson, arguably the most important for Canadian purposes, is about the "free rider." Free riders are firms and households that receive subsidies for doing things they would have done anyway. Say a firm or household was going to make an investment in energy efficiency. It has decided for whatever reason – cost-benefit analysis, moral concerns – to make the investment. Firms do this all the time: they study their costs and try to reduce where they can; they lower their energy costs as an input to production. Now along comes a government program offering a subsidy for what the firm or household planned or has even already done. The result is wonderful for the recipient: unexpected money. For government policy, the result is a bust. The subsidy program costs money but has not produced a change in behaviour. When governments or utility companies offer subsidies to those who claim to be improving energy efficiency, there is no easy way of ensuring that the money is being given only to those who were not already planning or implementing improvements. The money gets distributed, and free riders benefit.

Who knows how many free riders exist in a large subsidy program? Short of insisting on every recipient taking a lie-detector test, it's impossible to discern intentions. But the independent researchers noted above have examined this phenomenon, using a control group not subject to the policy and comparing the results with those from a group of subsidy recipients. Some of these studies estimate free ridership of at least 60 per cent.

Here's an everyday example of free ridership. A typical new fridge consumes about 500 kilowatts a year. A new energy-efficient fridge (suitably labelled, of course) consumes only 400 kilowatts, a savings of 20 per cent. If a government could only encourage consumers to choose the more energy-efficient fridge, it would appear that the energy and GHG savings would be substantial, at least in areas where electricity comes from burning fossil fuels.

Suppose that the government provides a rebate of $50 on purchases of the energy-efficient fridges in an effort to increase their market share. What would be likely to happen? First, as in all subsidy programs, the government's budget for the program will be limited. It cannot subsidize all fridge purchases, as this would be exorbitant. So it provides only enough money for perhaps 20 per cent of new fridge purchases. But fridges that are considered high-efficiency by government (Energy Star rated) already capture over 40 per cent of the sales in the market for new fridges, and this amount will grow over time, just as it did in previous decades. Almost all of the rebate could be captured by the people who would have bought efficient fridges anyway. The customers who would have bought the energy-efficient fridge anyway are free riders. The result for overall government policy of this kind of free ridership is obvious: the cost is very large relative to the effects.

Subsidy advocates insist, however, that each person who acquires that efficient fridge (or some other device) demonstrates to neighbours and friends the benefits of efficiency, creating a spillover effect. Each subsidy thereby hastens the adoption by others of something more efficient. Alas, independent research debunks that appealing idea, because energy efficiency is often more costly than it originally appears, which in turn limits the spillover effect. New technologies almost always have a higher risk of failure. Similarly, investments in more efficient devices require upfront capital expenditures, which can mean a long payback period, varying with the amount of energy savings versus initial investment (and forgone income from that investment).

A CRUEL PARADOX

There is also the cruel but observable tendency for energy efficiency to encourage greater energy use. More efficient technology can induce consumers to believe they are saving more per unit, and therefore to use more units. More efficient lighting, for example, will lower operating costs but can encourage people to leave lights on longer. Hybrid electric-gasoline vehicles encourage the development of vehicles with greater horsepower, thereby offsetting some fuel savings. This syndrome is called the "direct rebound effect." Although it is likely to be small for many end-users, it can be as high as 10 to 35 per cent in some cases.

Beyond the direct rebound effect lies the "mega-rebound," the link between energy efficiency innovations and expanded demand for related but new energy services. A recent study by Roger Fouquet and Peter Pearson on the history of lighting services and lighting use in the United Kingdom over several centuries found that efficiency gains in lighting had reduced its cost in 2000 to just 1/3,000 of its 1880 level, but total per capita use of lighting had increased 6,500 times.

CONSTRAINTS, POLICIES, ACTIONS

Achieving significant reductions of GHGs in Canada will require awareness of the constraints and intelligent, well-crafted policies that will lead to action, real action. For that to occur, we must rethink what Canada has been doing for so many years. It was illusory to imagine – when Jean Chrétien got Canada to ratify the Kyoto Protocol in 2002, merely six years before the beginning of the 2008–12 commitment period – that the country could make the dramatic domestic reductions required with the feeble policies that Liberal governments proposed. Those policies, as Canadians have come to understand, did not produce action, at least not the sort of action that leads to results. The atmosphere, it was decided without anybody in government describing the matter this way, would remain free as a trash bin for all. Targets could be announced and deemed achievable without any significant costs to consumers, taxpayers, or businesses. Having accepted these fundamental propositions, the government was left with only exhortation and subsidies, approaches that quite predictably failed.

Canadians now know, or ought to know, what policies will not work. If governments, environmentalists, and business will recognize that fact, we can finally focus on policies that will.

CHAPTER FIVE

Unhelpful Allies on Both Flanks

In 2003, Ontario Liberal leader Dalton McGuinty campaigned on a pledge to retire all of his province's coal-fired electricity plants by 2007. In November 2006, the premier changed course. Some plants would still be in operation as late as 2014. A reporter asked what lesson he had learned. "I have learned not to trust experts," he replied.

"Experts" abound in the environmental policy world, as they do in every area of public policy – or at least people who claim to be experts abound. Some try to remain as detached from political or interest group agendas as possible; others use their knowledge to push their agendas and advance their own interests. Environmental and business lobbyists are foremost among the "experts" who push and pull political actors, react to their declarations and policies, and try to influence public opinion so that it will influence government.

Federal, provincial, and municipal politicians of all political stripes have made various energy or environmental commitments, based on "expert" judgment, that turned out to have little or no effect. Politicians can make an announcement, as McGuinty did about closing coal-fired plants, and win hosannas of praise from environmental groups. Unfortunately, he had listened to their experts, who said the province's power needs could be filled by conservation and renewable energy such as solar and wind if the coal-fired plants closed.

On the other hand, politicians have often listened to business interests whose claims of looming economic catastrophe frightened them away from taking serious action against climate change. Environmentalists often praised the Chrétien and Martin governments for negotiating and ratifying Kyoto, making emissions reduction commitments, announcing climate change programs, and issuing "action plans," even though the record of these governments was weak; business interests lobbied the same governments, fearing Ottawa did not understand their concerns and might develop policies that would be too strong – fears that were never realized. Each side mobilized its experts to press its self-interest, always careful, as effective lobbyists are, to couch that self-interest as being in the broader national or even planetary interest. It turned out that both sides impeded progress, but for entirely different reasons.

A FIELD STUDY OF THE ENVIRONMENTALISTS

Environmentalists come from all walks of life and have a considerable diversity of views. This healthy diversity makes it too easy to offer a caricature of the "typical" environmentalist position on sustainable energy and climate change. The majority position, however, can be sketched out, with allowances for many deviations from the sketch.

Environmentalists dislike the big footprint created by large and growing energy appetites in wealthy countries. They sometimes find it difficult to accept that this appetite is spreading to areas of the developing world where citizens and governments want a better life. They deplore the rapid consumption of non-renewable resources, such as fossil fuels, since this consumption displays a disdain for coming generations that will need those resources. Some environmentalists offer their critique as part of a wider critique of capitalism, imperialism, and materialism; others, such as Canada's Green Party, call into question not the entire free-market system but just the incentives within it that produce deleterious effects on the environment, such as GHG emissions.

Environmentalists are naturally drawn to energy efficiency and renewable sources of energy. Greater efficiency reduces energy consumption, they believe, and less consumption means less energy production with its threats to the environment. They argue, too, that more renewables provide a sounder footing for our socio-economic system, because they will cushion us from the potentially cataclysmic economic and geopolitical consequences of rapidly depleting fossil fuel reserves. Energy efficiency and renewables are the twin pillars of a sustainable energy system. The solution to various environmental challenges, present and future, is assumed therefore to be the banishment of fossil fuels, or a steep reduction in their use. Also banished would be nuclear power, whose risks environmentalists consider overwhelming. Any new event – a series of violent storms, a power blackout, fewer polar bears, melting sea ice – will be quickly and even triumphantly interpreted as proof positive that humanity must stop or scale back the use of fossil fuels as quickly as possible.

Environmentalists also worry that society will not take environmental threats seriously until it is too late – and will not act if serious sacrifices are required. They have been right because until very recently in Canada, the public remained largely indifferent to the problem of climate change. Changing public opinion is hard, and they have had to steer a tricky course. On the one hand lies the danger that they will use such gloomy rhetoric that citizens will not listen. On the other, they must insist that energy efficiency and renewables are relatively inexpensive if they are to offer an alternative, reassuring future they can sell to fellow citizens.

The trouble is, as any scientist knows, that your objectivity is compromised if you hope too much for a certain outcome. So, not surprisingly, environmentalists are enamoured of research showing that households and businesses will profit from great improvements in energy efficiency. They are also easily convinced that Canada's Kyoto target can be achieved without requiring reallocation of funds from other priorities such as education, health care, income support, child care, essential infrastructure, public security, and national defence. It's just too unsettling for them to admit that meeting the Kyoto target for Canada is virtually impossible, even if it involves the presumed purchase of carbon credits from other countries.

The future of international carbon credits is highly uncertain. If Canada were to become a major purchaser in the Kyoto time frame, we could drive up the cost ourselves, especially if Europe is the only credible marketplace. If the price is in the $20-per-tonne range for 2008–12, buying 250 million tonnes in credits would cost Canada perhaps $5 billion a year, money that would have to come from other programs and would do absolutely nothing to reshape Canadian practices. Yet the hard-line environmentalists insist that this money must be spent – without indicating where it would come from, where it would go, and what good it would do in making the Canadian economy greener. They say Ottawa is running a surplus, so we should use it for climate change, without understanding the intensely competitive demands for that money from provinces, health care advocates, cities, aboriginal people, universities, and just about every institution in Canada – to say nothing of the tax reductions that a lot of Canadians want. In taking this position, the environmental lobby acts no differently than other lobbies that want money spent or action taken on the one issue, or set of priorities, that they have identified as urgent.

They saw a wonderful opportunity to press their case once the name "Kyoto" became conflated in the public's mind with concern about climate change. Put crudely – and it *was* put crudely in political discourse – being for climate change action meant being for meeting the country's Kyoto target. Cast doubt on the wisdom of this target, and you were labelled insensitive to climate change. As a result, politicians worked themselves into a lather of moralizing, with each opposition party professing greater fidelity to Kyoto than the other, such that they pushed through the House of Commons a Liberal MP's private member's bill in early 2007 requiring the government to produce a plan to meet the Kyoto target. Even when environmentalists privately admitted that meeting the Kyoto target was unwise, they could not stop demanding that the target be met, because they feared political pressure on the government for short-term measures would ease. That short-term focus was a shame, because serious debate ought to have been about what to do after the first phase of Kyoto – that is, post-2012 – rather than being fixated on an unrealistic target that Canada was not going to meet. But they feared any focus on the longer term would slow down any serious progress. There could be, in fairness, a mixture of purchasing credits and action at home, but domestic action within the Kyoto time frame would reduce only a fraction of the tens of millions of tonnes required, so billions of dollars would still have to be shipped overseas each year. Meeting Kyoto, given Canada's dismal record since the treaty was negotiated, would mean eliminating more than one-third of Canada's GHG emissions by 2012 – which would require sending crews to disconnect a third of our offices, homes, shops, and factories.

THE TOO-GOOD-TO-BE-TRUE SYNDROME

The flip side of the environmentalists' gloomy rhetoric is eloquent, incandescent hope that most people will see the light, as environmentalists define it, and shift to the low-footprint lifestyle they promote. They are like most people in their refusal to accept evidence that challenges their beliefs. Environmentalists will insist they have the greatest faith in science, and that science backs their every assertion, but present them with sound scientific research that contradicts their fundamental assumptions about the low cost and achievable possibilities of energy efficiency, or the inherent cleanliness of renewable forms of energy, and watch their blood rise.

The research that most interests environmentalists suggests great technological potential for energy efficiency. Their hero is the American guru Amory Lovins, who has made a career of preaching the ease and importance of energy efficiency gains. As Lovins told *Scientific American* in 2005, "Focusing on energy efficiency will do more to protect the Earth's climate – it will make business and consumers richer." Ralph Torrie and others, in a 2002 report for the David Suzuki Foundation, an institution that promotes low-footprint lifestyle as a key to "saving the planet," offered an unfounded assumption of energy efficiency profitability, saying, "Conservatively assuming a $15 cost saving for every gigajoule of energy productivity, the low carbon scenario would yield total annual cost savings rising to $30 billion by 2030, or a cumulative total of $200 billion between 2004 and 2030." There is an old rule that should be applied to predictions like this one: if something sounds too good to be true, it probably is.

If, for example, you focus on only the relative costs (investment and operating) of a light bulb, it would appear that the most energy-efficient device (which can be 75 per cent more efficient than its competitor) will also be the most economically efficient. By this measurement, it is estimated that throughout its economy Canada could profitably reduce its energy use by 35 per cent. There are "experts" who take their analysis to only this level, then write reports on the basis of which politicians such as Dalton McGuinty make promises about shutting coal plants. These profitable investments in energy efficiency, insist the "experts," can replace the one-quarter of Ontario's energy generation that comes from coal – a politician's win-win dream, the too-good-to-be-true syndrome.

An economy shifting to different patterns of energy use and placing a price on emissions will face economic costs for investments that would not have been required otherwise. Trying to reach the Kyoto target in a short space of time would impose costs on businesses and on the Canadian economy generally. Reasonable people could argue about how great the costs would be, and how to mitigate them. Reasonable people could also insist that if the proper regulations and incentives were in place, there would be future economic benefits. But environmentalists of the purest variety are not prepared to admit these realities. Instead, as crusaders for a cause, they paint wildly optimistic scenarios of nothing but economic gain.

Ralph Torrie said in October 2002, "We can systematically reduce greenhouse gas emissions in this country by up to 50 per cent by 2012, accumulating savings for Canadians that mount into the billions of dollars per year in saved energy costs [and] cleaning the air of our cities." Matthew Bramley from the Pembina Institute declared in June 2002, "Implementing Kyoto will actually boost competitiveness through increased energy efficiency, major new business opportunities in innovative low-GHG technologies, and by positioning Canada advantageously for the inevitable tightening of international restrictions on GHGs." Elizabeth May, then executive director of the Sierra Club and later leader of the Green Party, insisted the Kyoto targets were "totally achievable" and that Canada "could reduce emissions by 50 per cent by 2030 with existing technology."

THE BULB SHEDS LIGHT

Unfortunately, putative gains in energy efficiency require a more sophisticated analysis. How, for example, will the adoption of one energy-efficient device reduce potential gains from adopting another? More efficient light bulbs emit less heat into a room, which means that more heat will be required from the furnace. This does not mean energy-efficient light bulbs should not replace less efficient ones, but the net effect will not be quite as great as proponents argue.

Then there are the extra risks attributable to long-payback technologies, and the inability of some efficient technologies to substitute perfectly for conventional technologies. Again, take the simple light bulb. Yes, the ones based on more energy-efficient technology (fluorescents, for example) save operating costs, but not every consumer likes their different hue, tone, and other characteristics, compared with incandescent ones. Not everybody likes the hospital-like hue of some bulbs, and some prefer not to purchase them, even knowing they use less energy. Energy-efficient, and thus more expensive, technologies increase financial risks compared with older ones. You might buy an efficient bulb for $12, and if it lasts eight years, you get back the initial investment. If it breaks after one year, you lose $12. If the inefficient bulb breaks, you lose 89 cents.

And finally comes the reality that, human ingenuity and market demand being what they are, inventors keep producing new devices and services that apparently improve our lives but use more energy. This might begin to change – indeed, in some industries change is already at hand – so that ingenuity will be focused on even more energy-efficient ways of producing goods and services, but there will always be new products that consumers love that use energy, and lots of it.

Environmentalists argue that modern renewable technologies such as wind turbines, solar panels, biofuels, small hydro generators, and the like will eventually overtake fossil fuels, even without constraints or taxes on GHG emissions, or subsidies for the renewables. Over time, advocates insist, the cost of renewables will decline. Quite probably they will, as technologies improve. But the environmentalists' opposite assumption – that fossil fuel costs will inexorably rise as supplies decline – flies in the face of considerable evidence that the planet does harbour huge supplies of various unconventional forms of oil and natural gas, to say nothing of abundant coal. Fossil fuel production costs also benefit from technological innovation, just as renewables do.

IT'S AN EMISSIONS ISSUE, NOT AN ENERGY EFFICIENCY ISSUE

These beliefs, hopes, and myths of environmentalists can lead politicians astray. If, after all, energy efficiency and renewables are eventually going to be cheap, then strong compulsory policies today are not necessary. But here is the paradox: almost all the renewables do need subsidies to compete with fossil fuels. So governments head in two directions at once: subsidies, and exhortation to increase energy efficiency and use renewables, both of which allow governments to avoid the much more difficult and perhaps political dicey tasks of designing compulsory policies for industry and using price signals to guide consumers. Environmentalists demand that these tools be used, but they underestimate the costs of transition and overestimate the consumer take-up and the pricing advantages of the alternatives. They also invest far too much hope in the lifestyle changes they hope for or demand from Canadians. They cannot bear the prospect of Canadians actually using more energy, let alone the entire world using more, despite the fact that this will happen.

The truth is that Canadians are not, in one or two decades, going to abandon their desires for mobility, comfort, entertainment, and status. They might make some alterations, if given the right price signals in the market or provided with more energy-efficient products, but these will not be enough to drive down overall energy consumption. Environmentalists are all too often focused on energy, rather than emissions, in the belief that less energy use will inevitably bring down emissions; whereas the better way of approaching the GHG problem is to focus on emissions. Put another way, climate change is an emissions issue, not an energy efficiency issue. So the focus of intelligent policy should be on reducing emissions through much more effective thinking than the pious belief that energy efficiency will solve our GHG challenge, let alone the wishful thinking that Canadians will trade in their cars for bicycles, take Johnny or Sally and their equipment to hockey games on the bus, stop wanting conveniences and gadgets that improve their lives, hang their sheets to dry outside in –20-degree winter weather, and throw away their television sets and toasters.

BUSINESS AND THE RACE TO THE BOTTOM LINE

Like environmentalists, Canadian businesses defy compartmentalization. What links them, however, is the pursuit of bottom-line results. If they forget that focus, chief executive officers will be looking for another line of work, especially with shareholders, hedge fund managers, financial analysts, and competitors watching their every move. For many businesses, environmental concerns are "externalities" within which they operate in the drive for financial results. The environment is considered a constraint or fact of life, rather than the company's driving purpose. Sometimes governments impose constraints; sometimes nature does. Getting the balance right between making profits and dealing with the externality of the environment is not always easy, since companies and shareholders do not have the same interests as governments and citizens.

Climate change poses a particular challenge for businesses. A company that pollutes a river, spews particulates into the air, or deposits heaps of refuse leaves an observable by-product. Political pressure can then be mobilized against what has been observed. Greenhouse gases, by contrast, are invisible. That very invisibility is one reason why it has taken so long for the public to understand the link between GHGs and the environment – and why so little public pressure has been exerted on those companies that produce the GHGs to do something about them.

GHGs are invisible, but business complaints about the costs of reducing them, especially within the Kyoto time frame, have not been. For a long time, Canadians have seen and heard a steady stream of dramatic warnings from business. Adopt Kyoto, or impose serious measures on us to curb GHG emissions, and the economic effects will be devastating, corporate Canada insisted. Thousands of jobs will disappear. Companies will move to the United States or China. Canada's competitiveness will suffer. Investments will go elsewhere. The country's gross domestic product will slump. The currency will sink. This drumbeat from business has been relentless. Political forces opposed to serious action have picked up the beat and amplified it, as we have seen, dismissing Kyoto as a "socialist scheme." Sympathetic media commentators echo it.

So forceful is the message that Canadians could be excused for being afraid. They did not themselves understand global warming at first, and Canadian businesses and their lobby groups were filling their ears with dire predictions. It is much more reassuring for citizens, and governments, to listen to the corollary message from the business world that companies are already taking the necessary steps to deal with GHG emissions.

Much in vogue are declarations of "corporate social responsibility," a fine-sounding phrase that in reality rarely means that a company will do anything to reduce its economic returns in order to advance an environmental objective. The phrase more accurately means that firms will look for changes, slight or large, in practices that could contribute to environmental benefits, especially if these will not sacrifice economic performance or, better still, will improve that performance. What a business will do in its own self-interest, such as finding more efficient ways of using energy, thereby lowering input costs, can be trumpeted as an example of corporate social responsibility, useful for the marketing department or the chief executive officer's speeches.

THE COST OF IGNORING COSTS

Individual companies, and especially their associations and lobby groups, are right about one important aspect of an effective climate change policy: there will be economic costs. Environmentalists either deny this reality or minimize it. They are wrong. The dramatic reductions in GHG emissions that are required globally and by individual countries such as Canada will produce significant costs. Complying with Canada's Kyoto commitment would impose dramatic short-run costs, the apportioning of which could be done only through government decisions.

Business is right to draw attention to these costs, since firms will not willingly take them on, and many of them fear for their competitive positions. Well-designed compulsory policies could make the required technological transformations less traumatic; indeed, after the transformations, energy costs will not necessarily be significantly higher, and lifestyles need not be radically different.

If, however, governments and citizens do not accept that transition costs exist, they might be beguiled into believing that slight changes in practices can be a surrogate for more far-reaching changes. Such an attitude would do more harm than good, because it could lead us into thinking we are taking serious action on climate change when we are not. The refusal to acknowledge costs can also beguile governments into adopting policies that are deemed to be politically cost-free, such as our old friends exhortation and subsidy.

WHO'S AFRAID OF VOLUNTARISM?

Another kind of cost-free "solution" is voluntarism, a kind of lead-by-example philosophy, as noble in sentiment as it is ineffective in practice. Since voluntary measures cost little but can bring public relations benefits, companies have embraced the idea of voluntary action, often as a surrogate for real action. Voluntary acts by companies can be touted by them or their associations as demonstrating that problems are well in hand.

The Canadian Association of Petroleum Producers (CAPP) became the first industry association to sign on to the Voluntary Challenge and Registry, established by the Chrétien government to encourage companies to report their progress in reducing GHG emissions. And why not, since the VCR was entirely voluntary? Business, especially the oil and gas industry, touted the VCR as a means of drawing attention to the GHG challenge – not only for itself but for other parts of Canadian society – and of showing Canadians how diligently business was attacking the problem. Signing on was good public relations. It fits CAPP's Janus-faced approach to climate change: a reasonable public face pointing to progress through voluntary measures, but a vigorous campaign against any regulatory action.

CAPP faces an internal problem because the association represents so many interests in the oil and gas sector. Like many such organizations, it cannot advance positions beyond the lowest common denominator – what its most knuckle-dragging member companies would allow. The industry's bottom line therefore remained clear before and for many years after Kyoto: no regulations or serious economic measures forcing companies to reduce GHGs. Some of the larger companies recognized that, at least from a public relations perspective, denial of climate change and overt actions to delay any progress in the matter did not fit their worldwide image. Shell, British Petroleum, and Suncor, the oil sands mining company, have been among those companies wanting at the very least to admit that GHGs are a problem that has to be addressed. Others have taken a different tack.

THE CLIMATE CHANGE "HYPOTHESIS"

Some of CAPP's biggest members – EnCana, PetroCanada, Imperial Oil, Syncrude – denounced the science of global warming and warned of the most dire economic consequences if action were taken. ExxonMobil, Imperial's parent company, made debunking the science of global warming a worldwide corporate objective, financing scientists and lobby groups who disagree with the idea of global warming. Imperial Oil sent blizzards of information to Ottawa officials questioning the science of climate change and threatening cutbacks in investment and production.

"There are still too many uncertainties, too many theories chasing not enough facts, to commit to short-term policies with high economic and social costs, and uncertain benefits," said Robert Peterson, Imperial Oil's chairman. Almost three years after the start of the Kyoto talks, in 1998, Peterson still mouthed the ExxonMobil/Imperial line: "Is the world getting warmer or not, and if it is, why? Is global climate a real problem, and if so, then how serious? Does human activity have something to do with this or are we just looking at the normal variation in the world's climate?" A year later, Peterson's view remained that climate change was a "hypothesis," such that "I believe we need to learn much more before we will have a sound basis for policy actions."

Ron Brenneman, PetroCanada's CEO and a former Imperial executive, admitted that GHGs were an important public issue but preferred voluntary measures stretched over a long time. "We're not about to curtail growth because, I think, shareholders have invested in this company for shareholder value, not for us to solve a global problem," Brenneman explained. "I see fatal flaws in the arguments about global warming," said James Buckee, president of Talisman Energy. Talisman and Imperial Oil, working with lime producers, provided the financing for a conference in Ottawa of climate change deniers. When the federal government wrestled with the ratification of Kyoto in 2002, the business community created the Coalition for Responsible Energy Solutions, which took out newspaper advertisements warning of "severe disadvantage relative to the United States" and "higher prices and higher taxes." Gwyn Morgan, chief executive officer of EnCana, Canada's top energy company, wrote a public letter to Chrétien, saying, "It is my earnest submission that signing the Kyoto Protocol would go down in history as one of the most damaging international agreements ever signed by a Canadian prime minister. . . . Objective analysis of the facts shows that signing Kyoto would create a huge economic and environmental disadvantage for Canada."

The oil and gas sector, of course, can count on the Alberta Conservative government to reflect faithfully its concerns. What the sector counts upon is receiving public relations credit or, if any serious GHG reduction policy emerges, actual credits for improving its energy intensity – that is, reducing the amount of energy used to produce a given quantity of oil or natural gas. Such intensity improvements do occur at a rate of about 1 to 1.3 per cent a year, a logical outcome for businesses trying to reduce input costs. Intensity improvements, useful as they are, will be swamped by soaring volumes of fossil fuel production as the oil sands develop. Intensity improvements of this magnitude would only slow down the increase rather than reverse the trend of GHG emissions.

"APPROPRIATE FOR CANADIAN CIRCUMSTANCES"

CAPP's publicity just before the ratification of Kyoto was full of rhetoric camouflaging its opposition to serious action by the industry. The association said a proper "plan" for Canada needed to be "appropriate for Canadian circumstances." That meant recognizing that Canada is a big exporter whose economy is integrated with that of the United States and Mexico (which had not accepted reduction targets under the Kyoto Protocol) and that Canada's climate, geography, and population density make for high per-capita energy consumption. Other goals were that the plan be "achievable," "flexible," and "reconciled" with environmental protection, economic growth, employment, regional development, energy security, and aboriginal opportunity. As always, the industry reminded everyone that consumers, not producers, created most GHGs. In other words, the industry said it was eager to be part of a solution, but what it offered – modest energy intensity improvements – rhetorically covered the fact that it did not want to contribute much at all.

Later, the industry's position as expressed by CAPP advanced a little, when the association suggested that if companies did not meet intensity improvements targets, they would have to contribute to a fund that would be used to research more efficient energy technology. The energy industry, especially the oil and gas sector, has a shockingly poor research and development record. According to a study by the National Advisory Panel on Sustainable Development, Canadian industry spends, on average, 3.8 per cent of corporate revenues on R&D. The energy industry, however, spends a paltry 0.75 per cent. The oil and gas sector's record is even worse: a minuscule 0.36 per cent, less than one-tenth the Canadian average.

CAPP kept insisting the industry was making "significant" contributions by "working closely with the Alberta and federal governments," "reporting GHG emissions," and "participating in various international studies." When the Conservatives came to office in 2006, CAPP president Pierre Alvarez wrote to the ministers of environment and natural resources, reasserting the need for "policy certainty" based on "energy intensity improvement." But, he warned, there should be a "limit on industry's cost exposure to avoid undermining the international competitiveness of Canadian industry or harming economic growth." In other words, act if you must, but on essentially a business-as-usual basis. He added that policy should be "non-discriminatory" among sectors, a demand that reflected the industry's long-held determination to spread action across the economy. Alvarez also supported – who does not? – "development and adaptation of new technology" but warned that any measures should be "economically and administratively efficient," allowing for "flexibility in meeting emissions targets with minimum administrative burdens of implementation and compliance."

As the letter showed, the oil and gas industry's position had not changed in two decades. It wanted limited costs for itself, offered nothing beyond limited intensity improvements, and opposed anything that might undermine competitiveness, the assumption being that serious measures to combat GHGs would cause damage rather than create possible opportunities. Nowhere, at least in its public statements and policy positions, did the industry try to get ahead of the GHG policy curve. Rather, the industry fought a rearguard action against serious measures, correctly chastising governments for accepting unrealistic emissions targets to reduce GHGs but never offering a serious policy alternative that might move the country in that direction. Nor did the industry or other businesses worry about how high the costs might be of taking little if any action. Their analyses were therefore entirely one-sided, like looking at only half a corporate balance sheet.

The Coal Association of Canada took out full-page newspaper advertisements questioning the greenhouse effect. Roger Phillips, president of IPSCO, a Regina steelmaker, argued that Kyoto was little more than a nefarious attempt by Europeans to gain advantageous trade positions over North American business.

NO PLAN IS A GOOD PLAN

As it turned out, business need not have worried. The Chrétien and Martin governments never implemented a serious Kyoto plan. Government policy-making was sufficiently disorganized and political will was so lacking that reductions of GHG emissions never happened. The Chrétien government ratified the Kyoto treaty, but its 2002 plan and the Martin government's 2005 plan did not impose any burdens on business. Proposals were made for limits on large final emitters, but these targets were watered down, and the Martin government fell before anything was imposed on the LFEs.

Business had a valid complaint, from the Kyoto negotiations to the end of the Liberal era. Government policy-making was indeed often incoherent. The Kyoto target, had it been met, would have imposed considerable costs on business. Business was right that there never was a sensible plan, but business interests would not have welcomed one. Any sensible plan would have used economic tools and regulations. These would have been phased in to give business time to adjust. They would have provided certainty, maybe not of the kind business wanted, but certainty nonetheless. They would have looked to long-term change but begun the process of change during the first phase of Kyoto. But these were the very tools the business community most opposed. When business interests lamented the lack of a "plan," they were correct in the sense that those unveiled by the Chrétien and Martin governments would not have met Canada's ambitious Kyoto target, but wrong in the sense that any serious plan had to require major changes by industry. A plan without changes in the GHG-producing sectors of the economy was no plan at all.

Business was right to complain that Ottawa frequently sent confusing signals, but business itself often sent different signals to government that made coherent policy development difficult. Business would play up its public-spirited "corporate responsibility" and the virtues of voluntary compliance, presumably while understanding privately the limitation of this policy option, while at the same time repeatedly warning governments against any policy that would have the slightest impact on their costs. A more constructive approach would have been to get ahead of the GHG policy curve and help governments design innovative policies such as the carbon tax that Norway implemented in the mid-1990s. Norway built in special exemptions and government supports, to ensure that a significant foray into GHG reduction by industry would not compromise the competitive position of Norwegian firms, and indeed would help them become world leaders in zero-emissions fossil fuel use.

PREDICTING FINANCIAL DISASTER

Canadian Big Oil's campaign was endorsed by the Business Council on National Issues (BCNI), the big business lobby group now called the Canadian Council of Chief Executives. Its president, Tom d'Aquino, has never passed up an opportunity to underscore the disastrous economic consequences of the government fulfilling its Kyoto commitment. In the summer of 2002, the BCNI issued a major climate change paper that said reaching Canada's Kyoto goal was impossible, would undermine Canada's competitiveness, and would drive down investment, credit ratings, and the value of the Canadian dollar. Since Canada's population was growing, and Kyoto took no notice of this fact, Canada might be forced to lower immigration to meet its Kyoto target.

Instead, the BCNI said, the country needed a "made-in-Canada" strategy, the precise phrase used by Kyoto's political critics such as the Reform Party and later the Canadian Alliance and the Conservative Party. With a bow to Alberta, the BCNI insisted that no region should incur an undue share of the costs of GHG reduction. It urged governments to encourage economic growth, more jobs, rising personal incomes, and a stronger tax base, and it suggested that rather than target GHGs, Canada should focus on air quality in urban regions (shades of the Conservatives' campaign platform 13 years later). The BCNI quoted the well-known critic of the science of climate change Professor Richard Lindzen, of MIT, to buttress its case.

This made-in-Canada policy, advanced as an alternative to Kyoto ratification, predictably contained no impositions on business, except for industrial sectors negotiating non-binding "goals" with government. There should be "public education and involvement," investments in energy-efficient technologies, tax credits for renewables, new efficiency standards for buildings and appliances, research into sequestration, more public money for public transit, and more intelligent land use planning. Industry, said the BCNI, had made "very significant progress in improving energy efficiency" and was prepared to do more through better "energy intensity." That was all the corporate lobby was prepared to consider – what industry was already doing. For industry, therefore, the recommendation was largely the business-as-usual scenario; for consumers of energy and governments, however, the BCNI proposed action.

At the time of ratification in 2002, business remained universally hostile, as a government summary of consultations reported: "A wide range of industry stakeholders expressed the view that the Kyoto Protocol is badly flawed because it imposes absolute targets on only a small number of countries. . . . For Canada, this would mean higher costs and a shift of investment to other countries not bound by targets. . . . Many industry participants argued in favour of an alternative approach that would include longer time frames, less restrictive targets and greater harmonization with the U.S. approach."

The anti-Kyoto business lobbies proved successful. They did not prevent Canada from ratifying the Kyoto Protocol, but in every other way the lobbies prevailed throughout the Liberal years. Warnings against dire economic consequences resonated in the business press and among right-wing newspaper columnists.

When the fragile federal-provincial consensus cracked before and at Kyoto, industry formed a political pact of steel with Conservative premier Ralph Klein's government in Alberta and, later, with Conservative premier Mike Harris's pro-business, right-wing Ontario government. Having worked closely with Anne McLellan as natural resources minister, industry was delighted when Ralph Goodale from Saskatchewan replaced her, since Goodale represented an energy-producing province. When Herb Dhaliwal replaced Goodale, industry was delighted all over again, since Dhaliwal was a businessman who reflected industry's doubts about Kyoto.

DENY, DELAY, DEFLECT

Although some companies have vocally debunked the science behind climate change, others adopt an ostensibly less obstructionist position. They accept that human activity does contribute to climate change but argue that Kyoto is a flawed instrument, with excessively rigid timetables and no participation by major developing countries such as China and India. Before and after ratification, industry has insisted that taxes and regulations would impose insupportable burdens on the economy. The fossil fuel industry frequently publicizes the fact that the electric utility sector emits more carbon dioxide than any industrial sector.

When the fossil fuel industry cannot prevent action, it seeks delay. When action appears inevitable, it tries to deflect that action, except for voluntary measures the industry might take according to its own timetable. When action is indeed inevitable, industry seeks and obtains concessions. And if action is to be enforced, then industry prefers to measure progress through an "intensity index," whereby using less energy to produce more output is deemed a public good for which companies should receive credit – even if, as in the case of the oil sands, massively higher production of oil would swamp any improvements in intensity of energy use.

The record is tragically clear. Canada's climate change policies have resulted in a fossil fuel industry that in 2007 remains no more inconvenienced by serious GHG emissions controls than it was 20 years ago. Canadian governments have tried, unsuccessfully, to pursue emissions reductions while pushing economic growth in the oil and gas sector, including offering lucrative tax arrangements to promote exploration and development. With so much revenue flowing into federal coffers from oil and gas development, governments have always been reluctant to risk weakening their own finances today to help the environment later. Both Liberal and Conservative leaders have bragged about Canada's fossil fuel resources to foreign investors and consumers. Prime Minister Stephen Harper describes Canada as an "energy superpower" – without explaining how the country might also become a sustainable development superpower.

Emissions have fallen in some business sectors. The Canadian manufacturing sector, taken as a whole, has reduced total emissions, in part because some manufacturing industries have shrunk and in part because of critical process changes driven primarily by non-GHG reasons. Emissions from chemical production dropped by 28 per cent from 1990 to 2004, from pulp and paper production by 32 per cent, and in construction by 28 per cent. Iron and steel emissions rose only 1 per cent.

Some Canadian corporate executives have formed the Executive Forum on Climate Change, which released a statement in November 2005 endorsing the Kyoto Protocol process and urging governments to create a system after 2012 that would build upon the mechanisms in the first phase of Kyoto, such as emissions trading and a carbon market. "On an urgent basis," the executives called for investments in carbon dioxide capture and renewable energies. They accepted the scientific conclusions of the Intergovernmental Panel on Climate Change and noted that "Canada is particularly vulnerable to the impacts of climate change." "Deep reductions," they said, "are needed to prevent human interference with the climate system." This enlightened statement was signed by, among others, the presidents of Alcan, BC Hydro, Bombardier, Desjardins Group, DuPont, Falconbridge, the Home Depot, Power Corporation, Shell, and various insurance companies. Welcome as this statement is, most Canadian business leaders lag behind many American and European executives in speaking out about climate change.

CHAPTER SIX

Dion and Harper: New Leaders, Discredited Ideas

We all like to laugh at erroneous predictions, but sometimes errors bring sad consequences. All forecasts of complex systems are difficult, and invariably they are mistaken in some respects. Nobel Prize winner Wassily Leontief wryly underscored the reality when offering his 1981 forecast of population and energy consumption for the United Nations: "Regarding the projections, the only thing I am certain about is that they are wrong."

The best analysts struggle to predict how the world will change, even if they have a good understanding of what drives change. What is the appropriate response to the inevitability of at least some error? Stop forecasting? Project an extremely wide range of possibilities without assigning a stronger probability to any of them? The advantage of appearing to provide a forecast without actually offering one is nicely epitomized by the economist's quip to his students, "When making forecasts, give a number or a date, but never both." Confronting the reality that prediction is an inexact science, some might shrug and conclude that anything is conceivable – a recipe for not even trying to prepare for tomorrow by doing something today.

All sane humans carry in their mind at least some notion of how the world will unfold, or might unfold, and they base at least some of today's decisions upon that notion. We forecast everything from the change of a traffic light to our children's future and our own retirements. Business is always trying to forecast market changes. Politicians are always looking to anticipate the public mood. Foreign policy thinkers try to fathom where the world is heading or, in the case of global warming, by how much the earth is getting hotter and what that change will mean. People who say "I don't believe that human activity has produced climate change" are implicitly forecasting that future changes in climate will also not be caused by human activity.

We may scoff at some forecasts without saying that all of them are equally bad. People in the forecasting business spend a lot of time studying past predictions to understand why some were better than others. And there is wisdom in revising forecasts based on new knowledge, as the Intergovernmental Panel on Climate Change has been doing over its four reports forecasting the extent of global warming. Each one took account of new information, and the panel revised its predictions accordingly while narrowing the range of uncertainties.

Revising forecasts because of better information shows wisdom; revising forecasts to reflect one's biases shows – well, something else. Forecasts based on wishful thinking, for example, are certainly going to have the largest chance of being wrong. Wishful thinking has coloured a series of Canadian government forecasts about climate change – and it still infects a great deal of political debate about the challenge.

Consider the forecasts that the Canadian government has provided of greenhouse gas emissions. The next figure shows the forecasts of GHG emissions in 2010 that the federal government made in 1997, 1999, 2002, and 2006. The forecast lines depict how the government expected emissions to rise with "business as usual," in the absence of new policies. If at least some wishful thinking were not involved, you would expect the forecasts to err in both directions: too high or too low. The government, however, systematically erred on the low side. The government was wishing emissions would rise more slowly, thus making its ambitious rhetorical goals for reducing GHG emissions easier to achieve. Instead, emissions grew faster than anticipated, leaving the government with no choice but to correct each forecast and then offer another one, which in due course would also prove to be too optimistic.

Government Forecasts of Canadian GHG Emissions

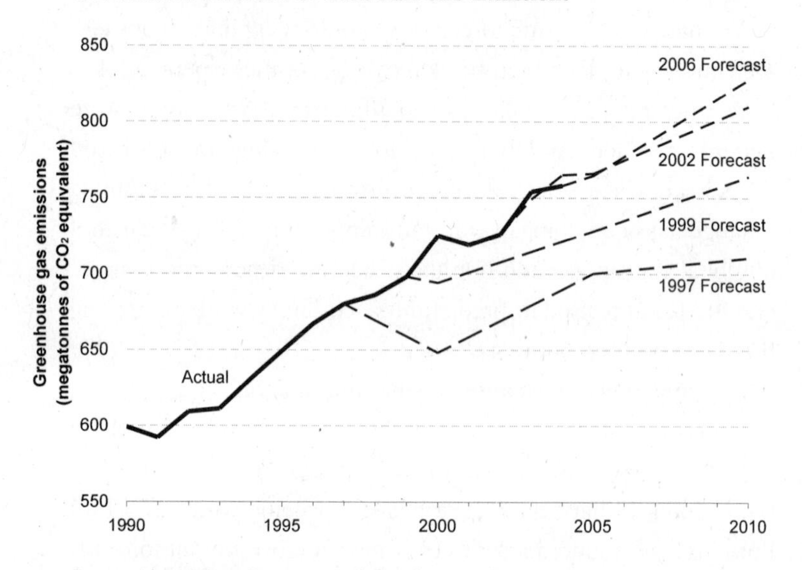

Source: Adapted from Government of Canada, 1997, "Canada's Second National Report on Climate Change"; Analysis and Modelling Group, National Climate Change Process, 1999, "Canada's Emissions Outlook: An Update"; Natural Resources Canada, 2002, "Canada's Emissions Forecast"; Natural Resources Canada, 2006, "Canada's Energy Outlook: The Reference Case."

Why was the Canadian government so consistently wrong? Wishful thinking might have caused it to consistently underestimate the growth of domestic GHG emissions under business-as-usual conditions. But some of the error arose from stronger-than-anticipated economic growth and more immigration throughout the period, both of which led to higher GHG emissions. Increased energy prices spurred the fossil fuel industry to new developments, such as the oil sands.

The Canadian government has layered one climate change policy over another. The record of past failures runs back almost two decades, starting with Brian Mulroney's last government, with the 1990 Green Plan that included $175 million for 24 GHG reduction policies, almost all of which emphasized providing information so that businesses and individuals could take action voluntarily. Then came Jean Chrétien's 1995 National Action Program on Climate Change, followed by Action Plan 2000 and the Climate Change Plan for Canada of 2002. The names changed, the policy approaches varied a bit, but the essential approach remained voluntarism and subsidies. Each plan forecast emissions reductions. The reverse occurred, as this figure shows.

Canadian GHG Emissions and GHG Reduction Policies

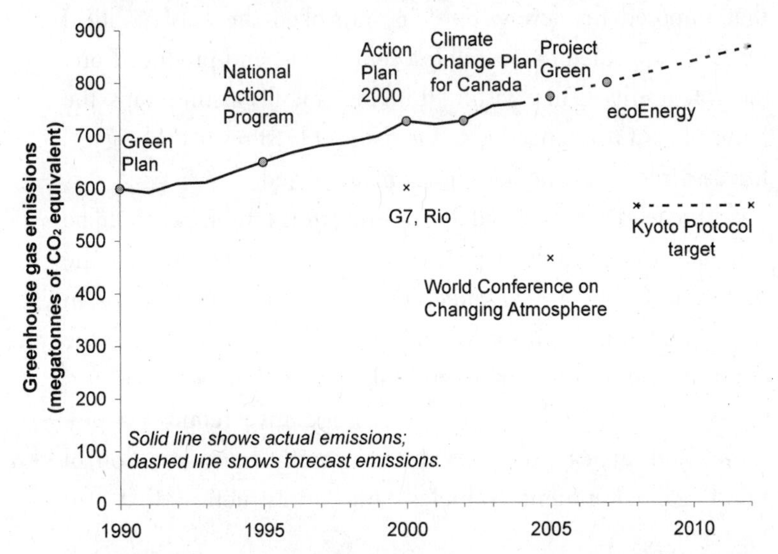

Source: Historic emissions from Environment Canada GHG Inventory, 2006; forecast from calculations by Mark Jaccard and Nic Rivers.

Key politicians engaged in wishful thinking, telling themselves and Canadians that targets could somehow be met without serious plans to meet them. The machinery of government got stuck in interdepartmental battles. Environmentalists oversold the possibilities of energy efficiency and the willingness – or the ability – of Canadians to change how they live, work, move, and keep themselves warm. Industry kept trying to destroy illusions about the ease and speed of reducing GHGs, but its own credibility was compromised by an unwillingness to present alternative policies that would have guaranteed substantial emissions reductions in a reasonable time frame. It did not help that some business leaders spent so much time and money delaying action by trying to convince Canadians about the "uncertainties" of climate change, even though this was not how these leaders would respond to comparable risks in their own businesses. And public opinion had not been mobilized to confront the challenge of GHG emissions.

All these factors led to policies that followed the paths of least resistance, voluntarism and subsidies, and so these were the paths that imposed the fewest burdens, provoked the least political reaction, appealed to the noblest sentiments, and produced predictable results. Call it wishful thinking or faulty assumptions: the paths of least resistance led to the paths of least results. The gains forecast from these policies never materialized.

Those who wish to avoid forecasting errors tomorrow would be well advised to study why forecasting errors occurred in the past. The Canadian evidence, and that from similar efforts in the United States to improve energy efficiency, suggests that the combination of voluntarism and subsidies will not work for GHG emissions reductions, because this imperative requires a profound and, at least in the short run, costly transformation of buildings, equipment, vehicles, and infrastructure that can occur only over a long time, with gradual but significant changes in prices and regulations.

Going forward, therefore, Canadians should examine what their political leaders are promising through the prism of past failures. Are politicians offering more voluntarism and subsidies, or variant versions of those policies? Canadians can be reasonably confident that these policies will not work and will nonetheless be quite costly.

Canadians now have new political leaders offering them new climate change approaches. Or are the approaches really new? Will these approaches work any better than previous ones? Urgency is in the political air. Each major party sounds more determined than the others to attack the GHG problem. Conservatives berate Liberals for having been so ineffective in office; Liberals decry Conservatives who opposed all the Liberal plans and now seem to mimic them. There are some days when Canadian politics looks like a St. Patrick's Day celebration, with politicians trying to outdo the others in sporting green.

THE NDP AND OTHER PARTIES

The New Democratic Party, to its credit, has been harping on climate change for years, and it elected Jack Layton in part because of his work in the field while a councillor in Toronto and head of the Federation of Canadian Municipalities. Environmental issues have sometimes bedevilled the NDP, because unions in the automobile, chemical, forestry, and mining sectors, among others, viewed environmental regulations as job-killers. Layton, however, defined his leadership as an environmentalist. His approach has usually showed strong preferences for subsidies, as indeed many NDP policies do because the party rarely believes in market solutions to problems. More recently, however, in the field of climate change, the NDP has come to promote market instruments to bring about desired social objectives on the climate change front. Layton has always stressed that he favours a cap-and-trade market system as part of the solution to GHG emissions, and that he wants to eliminate the tax breaks that spurred oil sands development. Unfortunately, the NDP got hung up on Kyoto, pressing policies to fit within the 2008–12 target timetable, without sufficient consideration for the practicality or cost of trying to meet Kyoto.

The Bloc Québécois, holier than thou as usual, insists that Quebec's environmental purity – a purity built exclusively on having been blessed with hydroelectric capacity instead of fossil fuels – is being held back by association with Canada, with its dirty oil sands, coal, and poor climate change record. With a new leader, Elizabeth May, the Green Party has grabbed some of the national political limelight and bids fair to increase its share of the popular vote in future elections. The three smaller parties certainly act as a prod, and in minority parliaments they have more influence over government decisions than in a majority parliament. Still, the Liberals and Conservatives are the parties with the big battalions. One of them will form the government. So it becomes more important to examine through the prism of past failures what these parties propose, to see whether finally Canada can begin to turn around its dismal record of limiting greenhouse gases.

THE DION EFFECT

Stéphane Dion won the Liberal leadership for many reasons, including the mistakes of his principal adversaries, Michael Ignatieff and Bob Rae, and because he identified himself with the environment. The man and the moment came together, for Dion had been Paul Martin's last environment minister, the author of Project Green in 2005, the president of the United

Nations conference on climate change in Montreal that same year, and someone who spoke passionately and often about the challenge of climate change. The Liberal Party was ready for the message. The party had a dreadful record on the issue, but in electing Dion, Liberals seemed to be seeking expiation for past sins. They also calculated that Canadians were getting greener in their interests, something the Conservatives by definition would not understand. The environment in general, and climate change in particular, would be especially arresting for young Canadians; the issue would cut across income groups, attract NDP and Green supporters, and reposition the Liberals as a party focused on the future.

Rhetoric and political calculations aside, Dion's actual policy prescriptions were not encouraging. His leadership campaign platform on the environment, a hefty 54 pages, was similar to the Project Green document he had released over a year earlier as environment minister. Since Dion remains wedded to these ideas (with one exception), it is reasonable to assume that they will define the party's approach for as long as he is leader. If so, as the following analysis demonstrates, Canadians should not hope for much.

Project Green, his policy chef-d'oeuvre, offered an emissions cap-and-tradable-permit policy for the largest industrial firms in Canada. These roughly 700 firms, together known as the large final emitters (LFEs), produce slightly less than half of Canada's GHG emissions. They include mining companies, major industrial plants, production and refining facilities for oil and gas, and electricity generation plants, and we have met them earlier in this book.

Under a cap-and-trade program, the government imposes a regulated overall national limit on emissions, then allows firms to buy and sell permits among themselves so that each firm meets its individual obligation. This is just the kind of market-oriented regulation that can be highly effective at forcing industries to make investments to lower GHG emissions – provided the regulation is properly designed, as it was not in Project Green.

First, the LFE program was based on intensity of emissions, not absolute emissions. For example, the target for a firm or a sector might be to reduce emissions per unit of production by 10 per cent by 2010, rather than to reduce overall emissions by 10 per cent. This approach would be a good start if the intensity reductions exceeded the growth of the output of the firm or sector, because then the plan would indeed lower absolute GHG emissions. With the LFEs, however, intensity requirements in the Kyoto time frame were set 12 per cent below the business-as-usual scenario, but this per-unit reduction would certainly be overwhelmed by anticipated industrial growth, especially in the oil sands sector.

Project Green also allowed industries to negotiate alternatives to GHG reductions in their own businesses. Firms had five choices: they could buy permits from other LFEs that had done better than their requirement; buy permits from government at $15 per tonne of CO_2; buy offset credits (a form of permit) from emissions reductions in other sectors of the Canadian economy outside the LFE system; buy GHG reduction credits from recognized international sources; or invest the money in a fund to research low-emissions technologies, instead of in permits.

The provisions had potential merit. The flexibility inherent in an emissions cap-and-trade system can reduce the policy's overall cost by encouraging firms with the cheapest way of reducing emissions to do the most. But all of the additional flexibility provisions meant that the probability of actual domestic reductions would diminish while the administrative complexity and cost would certainly rise. Even some industry representatives admitted in confidence that there was too much flexibility. Such a complicated system required extensive negotiations, and these had not concluded when the Liberal government fell. Even so, it is doubtful that the policy would have stimulated absolute GHG reductions within the Kyoto time frame of 2008–12.

The Project Green policy was aligned completely with that time frame, which was unrealistic because for major industries the investment time frame is much longer. Industries must consider economic returns over decades from major investments in plant and equipment. The policy provided no signals to LFE industries about what might be expected after 2008–12, thereby increasing the risk of making a short-term decision that would be proven wrong by subsequent developments.

What would have happened, however, if Dion's LFE policy had been implemented? To find out, we simulated this policy using the CIMS model. This is an energy-economy model that has been developed and applied over the last two decades by the Energy and Materials Research Group at Simon Fraser University to simulate environmental policies that influence technology choices by businesses and consumers. Independent researchers, governments, industries, utilities, and environmental groups throughout Canada and in other countries have used the model.

In CIMS the key energy-using technologies in our economy are tracked in terms of the amount they provide of energy services (square feet of home space heating or kilometres of car travel) or physical products (tonnes of steel or newsprint). New technologies are acquired to replace old equipment and satisfy the growing demands for services by households and for products by industry. Competing technologies, such as more and less efficient light bulbs, compete for market share of new investment for each service and product. CIMS simulates this competition using the best available cost estimates, plus relevant information about the particular preferences of businesses and households. (This information is important because one technology might look better in strict terms of cost and yet not gain as much market share because of the preferences of consumers. For example, if CIMS just looked at cost per kilometre travelled, every urban traveller would be using public transit almost all the time. But, of course, the private car has many advantages that are not reflected simply in cost per kilometre travelled.) In other words, CIMS incorporates the latest research on consumer and business preferences in this more comprehensive – and realistic – portrayal of the investment behaviour that all of us show in our workplace and domestic lives. Once all of the technologies are selected for a given time period, CIMS then shifts forward to the next time period. The current version of the model can cover time frames of from 10 to 50 years. During a simulation it sums up the evolution of society's capital stocks (buildings, equipment, infrastructure), energy use, and environmental outcomes, such as GHG emissions. While CIMS focuses on energy use, the model can be combined with forestry and agricultural research to provide total estimates of GHG emissions changes over time.

A simulation of Project Green using CIMS found that if it had been enacted in 2005, with permit prices capped at $15 a tonne, GHGs from the large final emitters would have dropped by 15 million tonnes by 2010, from where they otherwise would have been. That would have represented less than a 4 per cent reduction from sectors that are responsible for 50 per cent of Canada's emissions. Project Green gave the LFEs a 45-million-tonne reduction target, but 21 million tonnes would have come from the purchase of international permits from European countries and perhaps Russia – assuming the permits were available at less than $15 per tonne.

In other words, serious reductions from the LFE sector just could not have been squeezed into the Kyoto time frame. Continuing the

LFE policy that Dion proposed, if stretched over a much longer time frame, would have led to larger reductions, since firms would have made new investments if these produced reductions that were less expensive than buying international credits or other domestic offsets. By 2040, according to the CIMS model, the LFE policy would have reduced GHG emissions by about 43 million tonnes from the business-as-usual total. Most of the reduction would have come in the electricity and oil production sectors.

PROJECT GREEN AND THE AUTO INDUSTRY

Project Green also included a memorandum of understanding between the federal government and the automobile industry. The memorandum envisioned a reduction by 2010 of 5.3 million tonnes of automobile emissions, down from the business-as-usual annual 80 million tonnes of emissions from cars and light trucks. According to the government, this 5.3-million-tonne target corresponded roughly to a 25 per cent improvement in average new vehicle efficiency.

The contemplated GHG reductions were voluntary. Voluntary policies have a poorer record of compliance than mandatory ones, but the automobile industry fiercely opposed mandatory limits, in part because it was embroiled in legal action against the mandatory limits imposed by California and did not want to do anything in Canada to weaken its case. Twelve U.S. states adopted the California standard (together they account for a fifth of the U.S. auto market), and four of them have provisions that require them to align their policies if California toughens its standards even more. The Project Green memorandum allows automakers to terminate the agreement on 90 days' notice – which is presumably what they would do if emissions did not decline according to expectations.

While compliance within the Kyoto time frame is important, a much greater concern is the long-term effects of this policy approach. Without financial penalties or regulation of emissions, the manufacturers lack incentives to make vehicles with zero or near-zero emissions, or to readjust marketing strategies to promote more low-GHG vehicles rather than blitzing the airwaves with the usual advertisements for greater horsepower and more on-board devices (which increase fuel consumption). California, because of its regulations, leads the world in sales of low-GHG vehicles – without residents losing their love of the automobile. The only noticeable difference is the continuing rise in sales of Japanese vehicles, because these manufacturers have more rapidly embraced the new emphasis on low-emissions vehicles.

The same sort of effort, with regulations and emissions standards, is required from other parts of the transportation sector such as trucking, rail, shipping, and aviation. Project Green did not focus on these critical areas.

THE CHALLENGES OF WIND POWER

Project Green supported wind power with subsidies. The Wind Power Production Incentive and the Renewable Power Production Incentive provided subsidies of 1 cent per kilowatt hour for qualifying facilities. Also, changes to the tax rules in 2005 allowed faster depreciation of capital costs for renewable energy generation equipment such as wind turbines.

These subsidy policies are politically appealing and can appear to be effective. They may have little impact, however, on the adoption rate of renewable energy technologies, because the cost of renewables is so high. The cost of wind-generated electricity will vary, but it is rarely less than 6 cents per kilowatt hour and frequently above 10 cents per kilowatt hour in areas where electricity is needed, such as southern Ontario. This cost is high compared with the cost of electricity generated by fossil fuels such as natural gas and coal, if plants burning these are allowed to emit GHGs freely. And because wind is intermittent, wind power must be combined with investments in energy storage or generation facilities that can produce power when the wind is not blowing. Independent experts estimate that the full cost of energy storage from a natural-gas-driven turbine or a gas storage facility could increase the cost of wind-generated electricity by between 2 and 3 cents per kilowatt hour.

Even if Project Green's policy for large final emitters had proceeded, the cost of generating electricity from natural gas and coal would have remained highly competitive with the cost of producing electricity from renewables. This is why electric utilities in Alberta and Saskatchewan and elsewhere are still planning major coal-fired plants in spite of the Wind Power Production Incentive. The incentive is insufficient to overcome the cost advantages of fossil fuels as long as they are freely allowed to emit GHGs. This is also why Ontario intends to invest in natural-gas-fired plants producing at least 6,000 megawatts over the next decade if it keeps the commitment to close its coal-fired plants.

The federal subsidies have not driven the expansion of wind power, but provincial policies and voluntary decisions by electric utilities have. Most utilities in Canada have begun to emulate the example of TransAlta in Alberta, which started building facilities at Pincher Creek in the southwestern corner of the province, by committing to acquire some of their electricity at higher cost from wind power projects. Hydro-Québec is doing the same, supporting the investment in thousands of wind turbines. In 2002, British Columbia passed a clean energy requirement, obligating BC Hydro to meet 50 per cent of its new power needs from renewables and co-generation (the combined production of electricity and heat for industrial processes). In 2007, it upped the clean energy requirement to 90 per cent. Wind power investments have also benefited slightly from consumers who voluntarily pay extra for electricity. Consumers can opt to pay a higher rate, with the extra money transferred by their local utility to a company such as Bullfrog Power that produces electricity using only renewables.

Governments and utilities try to push wind power and take credit for wind power projects, but the fact remains that utility decision-makers in Canada and around the world still see combustion of coal and natural gas, without carbon capture and storage, as their cheapest electricity investment option. This reality explains why GHG emissions from the Canadian electricity sector have not fallen – a key point – and would not have fallen as a result of Project Green.

Wind power production might expand to 4,000 megawatts by 2010, but the Wind Power Production Incentive and the Conservative policy that replaced it should not take too much credit for this expansion. All or at least three-quarters of the expansion would have occurred anyway. And a 4,000-megawatt increase would reduce Canada's emissions by only 3 million tonnes from what they otherwise would be, fulfilling only about 1 per cent of Canada's Kyoto commitment. Dion's more recent claim of expansion to 10,000 megawatts by 2010 is a pipe dream rather than a serious policy, if he intends to bring it about by subsidies alone.

Given what utilities and provincial governments are contemplating, wind power in Canada could exceed 10,000 megawatts by 2050. Of this, not more than 2,500 megawatts, and probably much less, could be attributed to the incremental effect of the Wind Power Production Initiative. This attribution is a guesstimate because all wind power projects would be eligible for the subsidy, so it would be difficult to know for which ones the subsidy was decisive, and which enjoyed free rider status.

OTHER RENEWABLE SOURCES

Project Green also forecast that the Renewable Power Production Incentive would stimulate 1,500 additional megawatts of non-wind renewable energy, mostly from biomass and small hydro projects. A biomass combustion plant burns forestry waste (bark, sawdust, chips) or agricultural crop residues (straw, stocks) in a boiler to generate electricity from a conventional steam turbine. Biomass-burning plants or small hydroelectric plants run almost all the time, thereby generating much more electricity for a given installed capacity than wind turbines, which are dependent on erratic winds. As with the wind power subsidy, it is difficult to estimate the policy's effect. Canada has had a long history of developing biomass and small hydro projects without the Renewable Power Production Incentive. In each decade since 1950, Canada installed at least 1,500 megawatts of new biomass and small hydro capacity. All future projects will be eligible for the production incentive, which begs the question: How many of these projects would have proceeded without the subsidy?

If the Renewable Power Production Incentive were to continue until 2050, the CIMS simulation shows that it might cause an increase in capacity from renewables of about 500 megawatts over and above what would have occurred without the policy. There will therefore be more free riders than new ventures. Cost-effective biomass electricity generation in the pulp and paper sector is limited by the availability of wood residue (bark, chips) and combustible liquids from the recovery of chemicals in the pulping process. These will not increase appreciably in response to a subsidy of 1 cent per kilowatt hour. Small hydro generation is limited by the availability of quality sites close to transmission lines. Public opposition also attends some of these facilities. The subsidy will probably improve the attractiveness of some marginal projects, but that's all.

CO-GENERATION

Project Green assumed a 9-million-tonne GHG reduction by 2010 from expansion of co-generation. This is a decades-old technology in which steam from a boiler spins a conventional steam turbine to generate electricity while also providing heat for industrial processes or buildings. Achieving such a reduction would require about 16,500 megawatts of new co-generation facilities by 2010, over and above the amount of co-generation that would have occurred without the subsidy. This is highly unlikely.

Co-generation has long been competitive with coal, natural gas, hydro power, and nuclear generation of electricity in certain cases. Canada now has about 7,000 megawatts of co-generation electricity capacity. The sector had grown especially quickly before the new capital cost depreciation policy, and experts were predicting it would continue to grow as natural gas prices and supply concerns rose. Optimistically, Canada might install an additional 2,000 megawatts because of the accelerated depreciation, and this amount could grow to 9,000 megawatts by 2050. But the more realistic estimate of what the accelerated depreciation could bring by 2010 is 1 million tonnes of GHG reduction, not the 9 million tonnes claimed in Project Green. Again, there will be free riders, since all projects will be eligible for the subsidy.

DISSECTING THE CLIMATE CHANGE FUND

All these programs, however, paled in cost and sweep beside two others in Project Green: the Climate Change Fund and the Partnership Fund. These were the granddaddies of the subsidy approach. Like the rest of Project Green, they never were implemented, but since the subsidy concept is being resurrected by the Conservatives, it is worth dissecting them a little and estimating what they and the rest of Project Green might have done for lowering emissions.

The Climate Change Fund was given $1 billion to pay for emissions reductions through a "domestic offsets program" and "international credits." Canada could have counted both methods in its Kyoto accounting. International credits purchased by the Climate Change Fund could have been projects in the developing world or from industrialized countries that had ratified Kyoto. These credits were available in central and eastern European countries whose state-owned, polluting industries closed with the collapse of Communism. Since the resulting reductions occurred without effort by governments or industry, they are sometimes referred to as "hot air."

The domestic offsets program was supposed to pay for reductions that were outside the LFE program and incremental to what would have otherwise happened. Independent researchers studying past government and utility subsidy programs have estimated that free riders typically exceed 50 per cent and sometimes constitute as much as 80 per cent of beneficiaries. Moreover, because the domestic offsets system would have permitted retroactive credits for projects completed between 2000 and 2005, it would have been an administrative headache to figure out which projects were incremental and which would have happened anyway. Much of the Climate Change Fund, therefore, would have been spent on reductions that had already occurred or were going to occur without it, and some would have funded futile administrative efforts to figure this out.

THE PARTNERSHIP FUND

The other granddaddy subsidy program was the Partnership Fund, designed to beef up provincial or territorial projects to reduce emissions. This sounded excellent: governments working together. It had only a nominal, albeit potentially large, budget because the federal money would flow in response to initiatives from provinces and territories, and no one could predict how strong their interest would be. Its impact would also have been difficult to predict. For example, most business-as-usual forecasts for Canada already assumed some of Ontario's coal-fired plants would close, just as Premier McGuinty had promised after following the advice of "experts." But with the announcement of Project Green, Ontario immediately began negotiating for some federal support to replace power from the coal-fired plants with renewable energy. This switch was already provincial policy. Likewise, Ontario was going to link its grid more extensively to hydro sources in Manitoba and Quebec. So would money from the Partnership Fund actually have reduced emissions?

GRADING PROJECT GREEN

Many assumptions have to be made in estimating what might have been the total impact of Project Green. Some, notably the future price of carbon on domestic and international markets, are full of uncertainties. We assumed Ottawa would have spent $1 billion a year from 2008 to 2010, and would have kept on spending that kind of money until 2050. We assumed an average over time of $17 per tonne of CO_2 for domestic offsets and $8 per tonne for carbon credits on international markets. We further assumed major purchases of international credits by the Climate Change Fund.

And what do we find? That Project Green, as published in 2005, would have at best reduced Canada's domestic GHG emissions in 2010 by about 85 million tonnes from the business-as-usual path – far, far below the reduction of 270 million to 300 million tonnes required to meet Canada's Kyoto target domestically.

Canada would have had to purchase the remainder as international credits or investments in developing countries, in a very short time. The total cost of this purchase during 2008–12 could have been in the range of $25 billion, varying with the highly uncertain price of those permits.

Projections for Project Green extended over a longer period by the CIMS simulation produced predictable results. Even if fully implemented and sustained indefinitely, Project Green would have only slowed the growth in emissions. This figure shows the CIMS simulation of Canada's emissions if the country had followed the prescriptions of Project Green. The graph tells the story. This approach might work politically, but not in hitting emissions reduction targets.

Canadian GHG Emissions with Project Green Measures

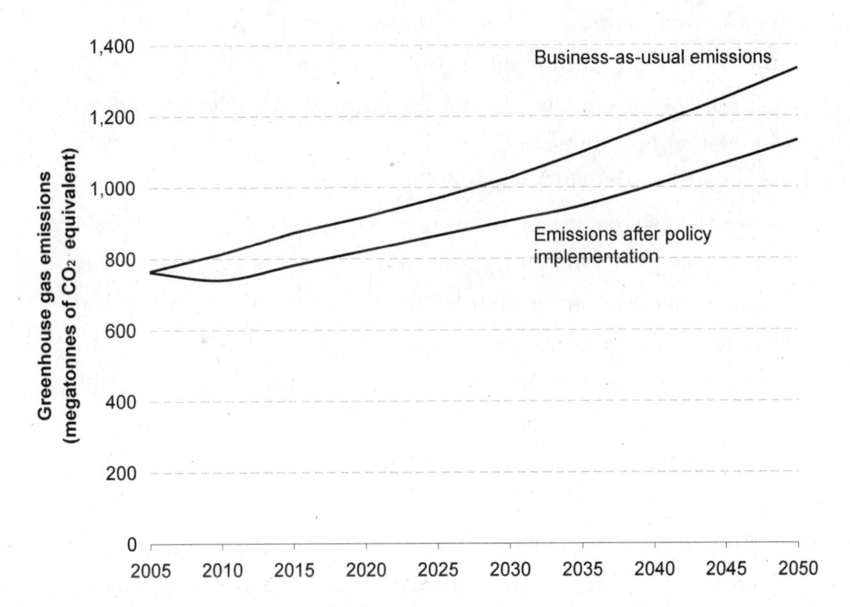

Source: Historic emissions from Environment Canada GHG Inventory, 2006; forecast from calculations by Mark Jaccard and Nic Rivers.

When campaigning for the Liberal leadership, Dion slightly modified Project Green. He still offered a wide range of subsidies: retrofit grants, green mortgages, tax writeoffs for energy-efficient vehicles, the One-Tonne Challenge, rebates for energy-efficient appliances. It had the same old large final emitters system, although he promised that after 2012 he would move from an intensity-based target to an absolute cap on emissions, a considerable advance. Although his new ideas were an improvement over Project Green, he generally remained wedded to the subsidy-and-information approach of the past.

As Liberal leader, Dion found the Conservatives contesting his primacy as a climate change campaigner. The Harper government was borrowing from previous Liberal policies – literally taking old, failed policies, adding or subtracting bits of money, changing a few administrative details, and giving the old programs new names. Rather effectively, the Conservatives were blunting Dion's advantage as the most committed advocate for action against climate change. In response, Dion ratcheted up Liberal commitments.

The Liberals' white paper of March 2007 calls for an absolute cap for LFEs at 6 per cent below their 1990 levels, with the program starting in 2008. On average, this would represent close to a 30 per cent reduction from current levels. Any emissions above the cap would be charged $20 per tonne of CO_2 in 2008, rising to $30 by 2012. The money would go into a "Green Investment Account" that the company could draw upon if it makes GHG abatement investments acceptable to a regulatory agency. Companies who get below their cap could sell the excess to other companies. They would also be able to pay for 25 per cent of their excess by Kyoto international mechanisms and would eventually have domestic offset opportunities, allowing them to subsidize GHG reduction activities by non-LFE parts of the Canadian economy. Again, there are no policies for the other 50 per cent of emissions caused by smaller companies.

This is the kind of thing politicians claim they will do when in opposition, much as Dalton McGuinty said he would close the Ontario coal-fired plants. One has to ask, Who were the experts Dion and the Liberals discussed this with? Because oil sands emissions are up much more than 100 per cent since 1990, starting next year oil sands companies will be hit with a tax of $20 per tonne of CO_2, rising to $30, on almost all their emissions. It is very unlikely that a new Liberal government would do this to this particular industry or to Alberta. In the nine years they governed following their Kyoto commitment, the Liberals were unable to levy even a $5 tax, or even the slightest cap on industrial emissions. By allowing so-called offsets, the Liberals are pursuing the same ineffective subsidy approach they have followed in the past. The only difference is that LFEs would provide the subsidy to non-LFE participants of the economy instead of government. The outcome would be no different.

With this policy in place, Canada would still come nowhere close to meeting its Kyoto commitment. Nothing much is happening to the 50 per cent of emissions outside of the LFE sector, yet these must fall by 30 per cent or so. In the short time frame of Kyoto, LFEs will not be able to reduce emissions significantly, and so would have no alternative but to pay the charge. Unless the Canadian government uses the money to buy hot air – a political impossibility – Canada will not come close to achieving Kyoto.

THE CONSERVATIVES' TAKE

The Conservatives had nothing good to say about Project Green when it was announced, or subsequently. Upon taking office, they promised a "made-in-Canada" approach that would be different, although they did not indicate how. They did, however, eliminate various programs that formed part of Project Green. Then, when the political pressure for action mounted in late 2006 and early 2007, they scrambled to produce something different. In many cases, however, they simply took Liberal programs, changed some details, budgets, and names, and reintroduced them.

Before that, the Conservatives' own climate change policies – or what were sold in part as climate change policies – were duds. They were not designed to be policy masterpieces, after all, but part of a platform that featured what might be called "tactile" promises, of tax breaks and direct payments to individuals.

A tax credit for public transit pass holders was one such policy, announced in a photo opportunity by Harper in front of a Vancouver bus during the campaign. The tax credit became part of the 2006 Conservative budget. It applied to monthly or annual pass holders. The credit would work out to a total tax rebate of about $190 a year for Toronto commuters or $380 for Montreal commuters, the difference being the cost of transit passes in the two cities.

A tax credit is a subsidy by another name, as it is a tax expenditure, whereby government forgoes revenue to support a particular activity. This subsidy would be a climate change success if, as Harper suggested, the tax credit got people to leave their cars at home and switch to public transit, since transit vehicles produce lower GHG emissions per person-kilometre.

Travellers, however, are very unlikely to switch in response to a modest decline in the price of transit. Current regular users of transit are the big winners. They are already purchasing passes and using transit systems, and they will now receive a tax credit for doing what they were already doing. They become the ultimate, literal free riders. The share of free riders among those getting the credit could reach over 95 per cent if, as expected, the measure induces only a few more people to take transit instead of drive. Using the CIMS model, we found that this policy will reduce emissions by only about 47 kilotonnes of CO_2 a year at an annual cost of $120 million – $2,000 per tonne of carbon dioxide. The transit subsidy, in other words, will be an expensive program with marginal benefits for climate change, whatever its political attraction, making it even worse than anything the Liberals did.

BIOFUELS FUEL FARM VOTES

The Conservatives also promised in the 2006 election to mandate 5 per cent average ethanol content in gasoline by 2010, and they delivered. They added, after negotiations with provinces, 2 per cent average biodiesel content in diesel fuel by 2012. But since ethanol cannot compete on a cost basis with gasoline, it needs a subsidy to be competitive. The Conservatives estimate their plan will cost at least $345 million.

These biofuels programs, although highly touted for their climate change benefits, are really more about finding new ways of subsidizing farmers. In the United States, the farm lobby from the corn states pushed corn-based ethanol very hard and secured in August 2005 a renewable fuels standard that injected billions of dollars of investment into the industry. In 2000, there were 54 ethanol plants in the United States producing 1.6 billion gallons. By 2006, the number of plants had grown to 97 and production to almost 4 billion gallons. The Renewable Fuels Association reported in 2006 that more than 150 other plants were in the planning stage. This explosion was sparked by regulations requiring a 2.8 per cent ethanol volume by 2006, plus income tax credits and additional biodiesel tax credits.

Canadian farmers, too, eyed corn-based or cellulosic ethanol made from grasses, wheat chaff, and corn stalks as a wonderful source of revenue. When announcing the Conservatives' plan, Agriculture Minister Chuck Strahl described one objective as being the creation of "new opportunities for our farmers and agricultural sector." The plan may be good for farm income, but whether it necessarily becomes good public policy for reducing GHG emissions is quite another matter.

The GHG impact of ethanol and biodiesel depends on how they are produced: what biomass they are coming from, and what kind of energy is being used to turn the biomass into fuel. The biomass can be corn, canola, or soybeans. The production of grain ethanol involves processing starches into sugars, which are then fermented into alcohol. The production of biodiesel involves the extraction of fats from oil seeds, followed by the breakdown of these fats to make biodiesel. Both refining processes need energy; refining grain ethanol is especially energy-intensive. Some analysts have concluded that producing grain ethanol requires more energy input than is released when ethanol is consumed in an engine. This should change, however, because a lot of effort is now being focused on finding more efficient ways of producing ethanol from cellulosic sources.

The problem remains that refineries producing biofuels are not guaranteed to help the climate, especially if the key policies are not in place. Ethanol plants could rely on electricity, but electricity is generally more costly than other fuels. Natural gas could be used, but its price has been volatile. So unless policies are in place that restrict GHG emissions or tax them, industries might decide to use coal to power biofuel refineries. This is happening in the United States, where coal is the energy source for many new biofuel refineries, and coal releases about twice as much in GHG emissions as natural gas. Professor Alex Farrell of the University of California at Berkeley has demonstrated that grain ethanol refined by coal-generated power actually produces more

GHGs than gasoline produced from conventional oil, even after accounting for the emissions required to extract crude oil and refine it into gasoline.

The GHG emissions reductions from biofuels are therefore dependent, in part, on which fuel is used to refine them. If coal is used, GHGs could actually rise; if other sources are used, the GHG reductions will be small, considering all the energy needed to harvest and ship the biomass to refineries. As an agricultural assistance policy, biofuels offer something new and perhaps better; as a GHG reduction policy, the value of biofuels is limited, to put it kindly, without an emissions cap or GHG tax.

THE "ECO" ECHO

The transit tax credit and the renewable fuels policy were campaign promises; what came later were policies the Conservatives designed on the hop as they scurried to meet mounting political pressure for environmental action. As with everything else the Harper government did, a tightly scripted thematic public relations campaign unfolded, starting with the labelling of every policy as "eco." So Canadians learned about the "ecoEnergy Technology Initiative," the "ecoEnergy Efficiency Initiative," the "ecoEnergy Renewable Initiative," and "Canada ecoTrust." The distinguishing characteristic of each program, and of the whole package, was their similarity to what the Liberals had proposed. "Canada's New Government," as the Conservatives rather ridiculously kept calling themselves even after they had been in office for more than a year, produced policies carrying different names but inspired by old thinking.

So, for example, the $300-million ecoEnergy Efficiency Initiative is very much like the Liberals' EnerGuide for Houses program, which the Conservatives axed. Both programs include a subsidy for energy retrofits of homes and small businesses. The size of the grants is similar, but the Conservative plan does not offer subsidies for the energy audits. Homeowners have to pay for these now, which will of course discourage some potential users of the retrofit subsidy. In any event, the Liberal program was expensive and had not done much to reduce GHGs. There is no way around the free-rider problem with either program, in that people who would have retrofitted their homes or business anyway would get the subsidy. The price of energy is a more important spur for people to retrofit than the existence of a subsidy, so for owners who decide to lower their energy input costs, the subsidy becomes a bonus payment rather than an incentive. The ecoEnergy Efficiency Initiative also contains public information programs about retrofitting. The program therefore followed precisely the information-and-subsidy model favoured by the Liberals.

The Canada ecoTrust was announced in February 2007, just before the Quebec election. Indeed, Harper announced with Premier Jean Charest beside him that Quebec would receive $350 million – more than the Quebec government had even demanded – of the $1.5 billion on offer. The timing and location of the announcement were part of the Conservatives' attempt to help Charest and themselves, by pitchforking federal money into the province. The substantive result of the ecoTrust program will be quite similar to that envisioned for the Liberals' Partnership Fund; indeed, the two programs are essentially the same. The federal government will subsidize GHG reduction projects undertaken by provinces, municipalities, or industries. Like the Partnership Fund, the ecoTrust program will be loaded with free riders and will therefore be expensive, but it will play very little part in decisions about investments in GHG reductions.

The ecoEnergy Renewable Initiative offers another $1.5 billion, this time to encourage renewable energy generation. Almost all of the money will subsidize qualifying renewable energy projects by 1 cent per kilowatt hour for the first 10 years. This program, too, is similar to Liberal programs that would not have had a large effect on GHG reductions, because the subsidy was not large enough to bridge the gap between conventional coal generation and renewables. And there will be free riders galore, because every province and utility company has already committed itself to renewable energy projects.

THE APRIL 2007 CONSERVATIVE PLAN

With the three opposition parties and virtually all environmental groups clamouring for Canada to meet its Kyoto target by 2008–12, the Harper government released a document in April 2007 outlining the cost of doing so. The opposition parties and environmentalists predictably attacked it, but five independent economists endorsed the findings: that to meet Kyoto would cause a recession in Canada, with gross domestic product dropping 6.7 per cent in 2008 and 7.2 per cent in 2009 from a business-as-usual emissions reduction scenario. Employment would actually fall, by 2.5 per cent in 2008 and 4.1 per cent in 2009. The report concluded, "All provinces and sectors would experience significant declines in economic activities." The shock of meeting Kyoto would be severe. Allowing for political overstatement, the document's conclusions reflected conventional economic analysis, even if the opposition parties and environmentalists did not want to understand the message. Their assertion that Canada's economy could survive serious action against climate change was correct – but only through a long-term adjustment, as opposed to this sort of sharp shock.

This document, and the publicity surrounding it, apparently did not convince the majority of Canadians, who kept telling pollsters they wanted Canada to achieve its Kyoto target within the 2008–12 time frame, a testament to the unfortunate conflation of climate change virtue with meeting Canada's Kyoto target. In part, this conflation reflected our basic confusion about the Kyoto target and the cost to the Canadian economy of meeting it, and also the heightened concern around the climate change issue, a concern the Conservatives had been scrambling to meet.

Months of "eco" announcements, flowing from a $1.9-billion climate change allocation in the 2006 budget and an additional $4.5 billion in the March 2007 budget, had helped to blunt criticisms that the Conservatives lacked climate change policies. But no complete accounting of the government's intentions could be made until it unveiled policies to the country's industrial sector. Therefore, the April 2007 announcement of a "regulatory framework for air emissions" was eagerly awaited and elaborately packaged – but it received a decidedly mixed reaction.

The government promised emissions regulations on large industrial facilities, mandatory emissions standards for passenger cars and trucks, strong regulations for household appliances, an emissions trading system for carbon, and a phase-out of incandescent light bulbs. The government pledged to reduce Canada's total GHG emissions to 20 per cent below their 2006 levels by 2020, en route to a 50 to 70 per cent reduction by 2050. Industrial emitters would be required to reduce the intensity of their GHG emissions (GHG per tonne of steel produced, for example) by 6 per cent a year from 2007 to 2010, and then by 2 per cent a year until 2015.

Put those pledges in context. Remember that Canada under the Chrétien Liberals had committed to reducing Canada's GHG emissions by 6 per cent from 1990 levels by 2008–12. Instead, by 2004 they had risen by 26 per cent. So by the time the Conservatives unveiled their policies, Canada had already missed its Kyoto target by at least 33 per cent. Along came the Conservatives and promised a 20 per cent reduction from 2006 levels, which may have sounded good – *but these were already 33 per cent above the initial Canadian commitment*. The Conservatives were therefore watering down the initial Canadian commitment, rather than promising in the next phase of Kyoto that Canada would make up for past failures *and* strive to become among the world leaders in combating climate change.

The Conservatives' approach looked at first blush to be based on a better mix of policies than anything we had seen before: mandatory emissions requirements for cars (the market-oriented regulation approach), a carbon emissions trading system, tighter regulations on manufactured goods (command and control). Finally, a government was going to use some of the economic and regulatory tools that could be successful over time in reducing GHG emissions. Alas, the fine print told another story, because the Conservatives' policies for large final emitters tracked previous Liberal efforts, complete with provisions that might more accurately be described as loopholes. And these loopholes would render the measures ineffective, eroding the likelihood of Canada meeting even the Conservatives' diluted targets for 2020.

The Conservatives, for example, announced their intention to negotiate mandatory vehicle efficiency standards. Vehicle efficiency regulations – just like industrial emissions intensity ones – may sound impressive, but they do not guarantee absolute reductions. If the number of vehicles and their rate of use grow fast enough – and they have grown very fast in the past decade – absolute emissions can rise, not fall. The same applies to more efficient appliances and light bulbs. In a growing economy, with increasing wealth and energy-consuming innovations, even substantial improvements in energy efficiency will not suffice.

The government insisted it would demand lower emissions, but at the same time it created loopholes that would allow industrial emitters to do something other than reducing emissions. This "flexibility" will allow at least some, if not many, emitters to undertake only limited in-house emissions reductions.

The large final emitters can claim free credit for reductions taken between 1992 and 2006 ("early action"). These reductions, however, will have to be verified by an independent review, a looming bureaucratic process. Fortunately, the regulations will cap this credit for "early action" at 15 million tonnes, but the rush to take advantage of the credit will mean that industrial emissions intensity will not actually fall by the percentages claimed by the government.

The LFEs can also subsidize others to reduce emissions, in Canada or abroad. These subsidies for others to reduce (in lieu of in-house emissions cuts) can be funnelled elsewhere in Canada, almost certainly to those not in the LFE sectors, or overseas (capped at 10 per cent) through the Clean Development Mechanism of the Kyoto Protocol; the largest recipient of these funds to date has been China. These so-called offsets are likely to appear quite cheap, but that is in part because payments are being made to people or companies for actions they were already planning, in many cases.

The LFEs will also have the option over the next decade of simply paying government for the significant amount by which their emissions exceed their intensity requirements. This approach, long promoted by the oil and gas industry, will create a "technology fund." Disbursements from this fund will be spent on projects that might reduce future emissions, say, by building a pipeline to transmit CO_2 for capture and storage – again, an idea that the industry and Alberta government have been talking about but doing nothing about for at least a decade – or building a long-distance electricity transmission line for sending hydro power to neighbouring provinces. In 2010, firms can meet 70 per cent of their obligations to reduce emissions by making such payments – a payment of $15 a tonne of carbon, exactly the price the Chrétien government had promised after the ratification of Kyoto. The share of their obligation that firms can pay in this fashion declines to zero by 2018. The per-tonne payment rises to $20 by 2013.

This arrangement provides short-term relief to industry. It can pay instead of reduce, the hope being that the payments will eventually produce infrastructure and technological breakthroughs that will lower the cost of future emissions reductions. Industries where no technologies exist to reduce emissions will be exempted. These emissions are about 10 per cent of the total. It means, however, fewer reductions than claimed.

As with the Liberals' Project Green, we tested these observations using the CIMS model. We wanted to test the assertion that these policies from Canada's self-described "New Government" would reduce emissions by 20 per cent in 2020 from 2006 levels. We also wanted to simulate the longer-term effect of continuing with these policies until mid-century, when, according to the Conservatives' plan, emissions will be reduced by 50 to 70 per cent.

We had to make assumptions consistent with leading research. So, for example, we assumed that subsidy programs, including offsets purchased by large final emitters, would be only 50 per cent effective. We assumed that funds generated by the technology fund would be only 50 per cent effective in reducing emissions. We assumed that government would require emissions intensity to improve at about 2 per cent a year after 2015, although the Conservative policy is silent on this point, and the flexibility provisions would remain unchanged. And we assumed that industries would get the maximum possible benefit for process-related emissions and early actions.

The following figure provides the result. Under these reasonable assumptions, the Conservative policies will not lead to a 20 per cent drop in emissions by 2020. The more likely outcome is that emissions, far from declining, will actually continue to rise, albeit at a smaller rate than in recent years. Conservatives, like Liberals before them, have a cap on industrial emissions that is full of holes, and they have no cap or tax on emissions arising from the other half of the economy. As a result, emissions will keep rising, and Canada will keep failing.

Canadian GHG Emissions with Conservative Measures

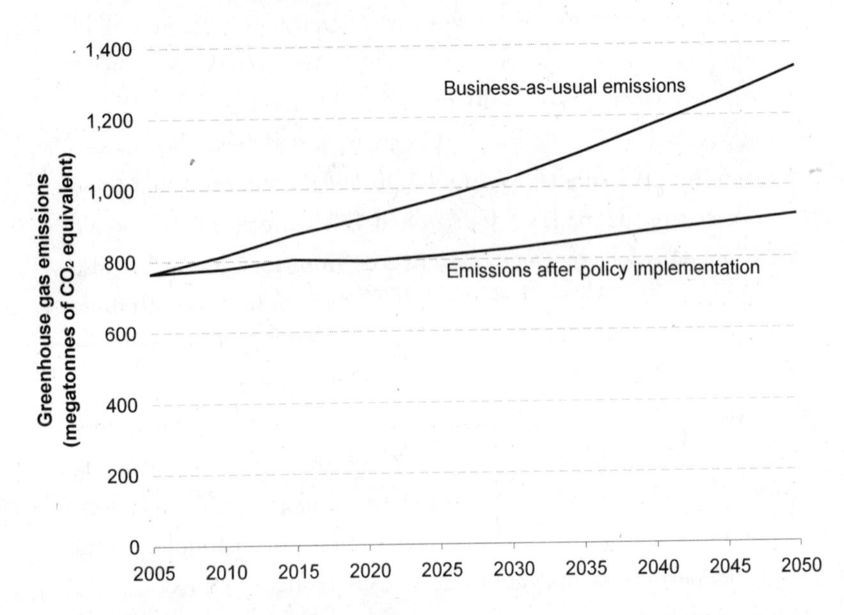

Source: Historic emissions from Environment Canada GHG Inventory, 2006; forecast from calculations by Mark Jaccard and Nic Rivers.

CHAPTER SEVEN

What We Should Do

Canadians ought to know by now what does not work in lowering greenhouse gas emissions. We have witnessed two decades of information and subsidy policies from our leaders – policies that continue to inspire the Harper government. The numbers do not lie: Canada's record of greenhouse gas emissions is appalling, and the information and subsidy policies of yesterday and today will not materially change that record. Worse, as governments develop increasingly expensive policy initiatives, such as the Harper government's "eco" policies throwing billions of dollars into all kinds of programs, the cost of failure grows in wasted money and time. In the scathing words of Johanne Gélinas, Canada's commissioner of the environment and sustainable development at the time, describing the history of Canada's climate policies, "On the whole, the government's response to climate change is not a good story. . . . Our audits revealed inadequate leadership, planning and performance." Gélinas further noted that to address climate change effectively, a "massive scale-up in efforts is needed" by the federal government.

A massive scale-up is indeed what Canada needs to reduce GHG emissions, but not just any collection of policies, however massive, will suffice. Politicians, when they think of doing something "massive," instinctively think of spending more taxpayers' dollars. This instinct leads to the politically attractive option of crisscrossing the country announcing funding for this or that special project, as Canadians observed when the Conservatives rolled out their "eco" projects at a series of photo-opportunity announcements by Prime Minister Stephen Harper, with ministers playing their assigned roles in the background like bobblehead dolls.

Photo ops and eye-popping financial commitments seem irresistible to politicians, even though most of the announcements miss the target of what must be done. There is no silver bullet to reduce GHGs, but one cardinal principle stands out: *The only way Canada can lower emissions appreciably over the coming decades – and this will be a decades-long challenge – is to design and implement either charges on emissions or regulations on emissions or technologies, or a mixture of both. We need economic tools and/or regulations to get the job done. There is no effective alternative.* Until Canadians and their governments understand this truth, we will continue to squander money, waste time, pursue variations of failed policies, and make scant progress. We might even continue to go backwards. We need, in other words, to stop digging in the same hole.

Canadians want answers, and if those on offer for so many years cannot suffice, which ones will? This chapter and the next one aim to provide a credible set of answers, illustrating the kind of policies governments can adopt that will lead to success. We will apply the CIMS model to our own policies to show why they will work much better over time than the Liberal and Conservative plans examined earlier.

Bear in mind two points in all that follows, and in everything you hear in public discussion of GHG emissions. First, *successful policies will require decades to produce substantial reductions in GHG emissions*. But we need to start implementing such policies as soon as possible, because the more time we fritter away pursuing failed policies, the greater the subsequent challenge of reducing GHG emissions. Second, while the specific design of GHG policies obviously matters to individuals, regions, and industries, *the bedrock idea of any approach must be that unfettered, cost-free dumping of GHG emissions into the atmosphere will no longer be permitted. The atmosphere can no longer be considered a carbon dump.*

EFFECTIVE POLICIES

Long-term, effective GHG reduction policies must be designed with several features in mind. For starters, policies that impose compulsory requirements are essential, because they provide the surest guarantee of action. These policies include traditional command-and-control regulations, taxes on GHGs, or market-oriented regulations, such as emissions caps with tradable permits. An approach that works may involve all three tools. For example, the United Kingdom, one of the world leaders in reducing GHGs, has all three: command-and-control regulations, a carbon tax, and an emissions cap-and-trade system for large industrial plants.

To be effective, these policies must minimize their drag on the Canadian economy. Using market-based policies is the best way of accomplishing this objective, because taxes and emissions trading schemes send the same financial signals to all firms and individuals, encouraging them to seek the lowest-cost actions that lead in turn to lower overall costs for society.

THE NATURAL LIFESPAN OF OBJECTS

Policies should also be coordinated with the rate of investment in new or renewed equipment and buildings. One of the hardest things to keep straight in discussing sensible GHG reduction policies is this central point: physical things have natural lives, which must be taken into account. Cars generally do not give out after three years; equipment will run for years; buildings are constructed to last for decades. Economists call this the natural rate of capital stock turnover. Policies that force the premature replacement of equipment and buildings that are still in prime condition cause significant economic costs.

Consider an electricity generating company that built a large coal-fired generating plant in 1995 at a cost of $1.5 billion. It complied with all the environmental regulations in 1995. These allowed the plant to emit about 2.5 million tonnes of CO_2 each year. The plant was designed to last 50 years and to generate electricity at a cost of 6 cents per kilowatt hour. If, as part of a government's stringent climate change policy, a GHG tax of $100 per tonne were imposed, the cost of generating electricity would rise to 15 cents per kilowatt hour, more than double the original cost. If, because of the tax, it therefore became cheaper to close the plant prematurely and replace it with a brand-new nuclear power station or hydroelectricity power dam, these massive investments would be an extra cost to the economy. The sudden imposition of this kind of carbon tax would create similar "premature obsolescence" for industrial plants, buildings, and vehicles throughout the country. Canadians would be forced to reallocate investment capital from other areas such as health care, education, security, infrastructure, and productivity improvements.

Policies must therefore focus on *new* investments. Sound GHG emissions policy should ensure that a coal-fired plant of the kind constructed in 1995 would not be built today – since a new plant could last until 2050 or even 2075. That is why it is important to start putting in place the new regulations and economic policies as soon as possible. To give businesses and individuals enough lead time, these new approaches should be phased in, being made more stringent as time goes on. Using these tools, GHG reduction policy can begin a transformation of buildings and equipment that will seem slow at first but will gradually percolate through the entire capital stock. It would be difficult and expensive to eliminate existing polluting buildings, plants, and vehicles, but by acting now and gradually toughening their approach, governments can influence yearly investment decisions of firms and individuals.

Not all governments seem to have gotten this message, although to its credit British Columbia, for example, adopted a policy in 2007 of not allowing new coal-fired electricity plants unless they deploy carbon capture and storage. British Columbia's policy is cleverly directed at new investments, not existing plants. As of mid-2007, the Alberta government of Premier Ed Stelmach, much in thrall to the province's fossil fuel industries, has not adopted the same approach for future coal-fired plants. It should be the case everywhere that *new* plants will not be permitted unless they are outfitted with technology that does not permit GHG emissions.

Climate change policies must be inspired by both effectiveness and economic efficiency. Sensible politicians, who have to explain and sell policies to voters, must also focus on two other factors: administrative feasibility and political acceptability.

HOW EASY IS IT TO ADMINISTER?

A policy's administrative feasibility depends on how easy it is to set up, how accessible it is to the citizen who wants to understand it, and what it costs the government and the individuals or firms that must comply. Poor administration can erode any program's good intentions, to say nothing of public support. The gun registry is perhaps the best example in modern times of administrative snafus and exploding costs destroying a program's credibility. The inefficiencies of the refugee determination system – the long delays, the multilayered appeals, the deportation failures – have eroded support in many quarters for that program.

So how simple would effective GHG reduction policies be? A GHG tax is relatively easy to administer, at least for CO_2, because it is tied to the amount of fossil fuels consumed. An emissions cap-and-trade system, by contrast, could be administratively complex if it were applied at the level of final consumers. It would be relatively easy, however, if limited to the roughly 700 large industrial plants that account for about half of Canada's overall emissions.

POLITICAL ACCEPTABILITY CONCERNS POLITICIANS

Political acceptability is even more important. For a long time, the politics of climate change in Canada seemed daunting. Liberal governments, while accepting aggressive, unrealistic GHG reduction targets, never thought Canadians believed that vigorous action was needed and therefore expended very little political capital in persuading them otherwise. The Harper Conservatives thought so little about climate change – and believed Canadians shared their indifference – that they put only two sentences about the problem in their 2005 campaign document. A stunning change came over Canadian attitudes in 2006, such that pollsters discovered climate change eclipsing the Canadian holy of holies, health care, at the top of the list of national concerns. As a result, policies that might have seemed politically unacceptable just a few years ago might now be welcomed, or at least not opposed. There have been upsurges in environmental concerns, including about climate change, in the past, but they have not been sustained. Now, however, the evidence of past failures, the mounting evidence of damage to the environment in Canada and abroad, and the overwhelming scientific consensus have seized the public's attention.

Of course, policies that concentrate costs on specific industries or regions are less likely to succeed, since in such a diverse country with a malignant sense of regional envy, the aggrieved region (or industry important to a given region) will cry foul, build a bonfire of outrage based on historic grievances, and generally make life miserable for the government in Ottawa. And no one should underestimate the political difficulties of imposing GHG taxes, since high taxes of any kind, anywhere, are prime targets of outrage – even when, as in the case of the goods and services tax, they replace other taxes rather than increasing the overall tax burden.

Thus far in Canada's discouraging climate change saga, political acceptability, or rather the perception of political acceptability, has driven government policies. Policies that were deemed acceptable – information and subsidies – were also the least effective ones. Now, with public opinion having shifted and somewhat greater public understanding being shown about climate change, a reasonable chance exists that serious policies can meet the test of political acceptability.

No policy will perform perfectly on this test, or on effectiveness, economic efficiency, and administrative feasibility. Trade-offs will always be required, and not all of them will be easy. Let's take an example from the world of automobiles. Stringent regulations permitting only the sale of high-energy-efficiency vehicles would be effective, but they would not be popular with manufacturers who derive much of their profits from larger and more fuel-intensive vehicles such as SUVs and trucks. Even the best policies – such as those outlined in this chapter – will not get top marks every time, but they will not score poorly on any of the crucial four tests.

IN PRAISE OF COMPULSORY POLICIES

Voluntary policies such as information programs, labelling of equipment, moral suasion campaigns, and non-compulsory agreements with industry will not be effective in producing deep reductions in GHG emissions. Canadians ought to know this, because their governments have been trying these policies for many years, with no overall positive results in the nation's GHG emissions record. As the Organisation for Economic Co-operation and Development noted after reviewing voluntary programs in various member countries, "While environmental targets of most – but not all – voluntary approaches seem to have been met, there are only a few cases where such approaches have been found to contribute to environmental improvements significantly different from what would have happened anyway."

Subsidies, too, are ineffective because, as we have seen, they tend to have high rates of free riders. Subsidies don't bring about permanent behavioural change because they are not large enough – or, if they are, the cost to the treasury is enormous relative to the change. Information and subsidy programs on their own are inadequate. They can, however, provide support for more compulsory policies of the kind that are essential for reducing GHG emissions.

The bottom line is something politicians just don't want to discuss, perhaps fearing the political reaction: if Canada is to reduce emissions significantly over the coming decades, policies must constrain by regulation or financial penalty the free dumping of GHGs into the atmosphere. Polluting behaviour must have a price, not in moral opprobrium but in financial terms. In a free-market society, to throw away the price mechanism is to discard a basic determinant of behaviour, but this is how governments have framed policy for two decades. No wonder their policies have not worked.

As the economics department of the TD Bank Financial Group concluded in a 2007 report, "Most economists, including ourselves, believe that any injury inflicted on Canadian jobs, incomes and competitiveness can be mitigated through reliance upon market-based policies that change the price structure to pollution. Doing so serves two purposes. It ensures that polluters pay for the social cost of their actions. And, it alters behaviour when the price for pollution becomes steep. Polluters will seek alternatives, thereby spurring innovation and reducing the need for further, more intrusive and costly environmental policies."

Compulsory policies can and will be effective. They will face, however, their own challenges of economic efficiency, political acceptability, and administrative feasibility. As we have seen in Chapter 4, the three major categories of these compulsory policies are command-and-control regulations; emissions charges or taxes; and market-oriented regulations that cap emissions or force the gradual adoption of particular types of vehicles, buildings, or electricity generating plants.

COMMAND AND CONTROL

Economists fear that command-and-control regulations will be economically inefficient. These policies, however, remain popular in certain circumstances in many parts of the world, and we already use command and control in Canada. Through the Canadian Energy Efficiency Act, the federal government sets standards for minimum efficiency of most large household appliances such as refrigerators, washing machines, and freezers. Through building codes, some provinces require minimum amounts of insulation and furnace efficiencies, and so on. In this way, command-and-control regulations at least eliminate the least desirable technologies. Indeed, by constantly tightening these sorts of regulations, every decade or so, governments can gradually phase out the least efficient models currently available.

EMISSIONS CHARGES AND CARBON TAXES

Economists leery about command-and-control approaches feel more comfortable with taxes. They don't have to get elected, of course. Prime Minister Chrétien ruled out any carbon taxes soon after Canada signed the Kyoto Protocol, and no Canadian politician has promoted them since. In fact, the Reform, Canadian Alliance, and Conservative parties frequently accused the Liberals, notwithstanding Chrétien's statement, of harbouring a hidden agenda to impose such taxes. Michael Ignatieff raised the possibility of considering a carbon tax while running for the Liberal leadership in 2006, and other candidates immediately denounced him for the heresy. The short political history of carbon taxes in Canada shows the idea to be one that is worth discussing but is not taken seriously in the political arena because of fear and opportunism, media hysteria, and equally hysterical business denunciations.

Set those sentiments aside and consider a GHG tax. Scandinavian countries have used them for years, Norway having introduced one in 1991. In Britain, where GHG emissions have fallen 7 per cent since 1995 despite 25 per cent economic growth, a report from the cabinet secretariat said in early 2007, "If industry and consumers are made to face the true damage cost of each tonne of carbon they emit, they will produce the socially optimal amount; that is, up to the point where abatement becomes more expensive than the damage caused by CO_2. Carbon trading schemes and taxes are both means of bringing those costs to bear on polluters." A GHG tax would be set initially at a modest level, but would gradually rise. Eventually, it would reach a price at which zero-emissions technologies would frequently be the most economic option for business and consumers. A tax that eventually reaches $100 to $150 per tonne would produce this outcome. The ultimate effect would be to increase the annual cost of energy services to households by between 25 and 50 per cent a year by 2050, which sounds terrible until you realize that it means an annual increase of only less than 1 per cent.

This increase represents the extra cost of energy efficiency, fuel switching, and GHG capture and storage in whatever combination these actions would be chosen by governments, businesses, and consumers. By 2050, with this sort of tax and the changes it would bring, emissions would be half the projected levels if there is no tax. The costs of energy equipment (such as energy-efficient light bulbs) plus the running costs of energy would take 8 to 9 per cent of a typical family's budget, instead of 6 per cent today. The higher the tax, the greater the emissions reduction, but also the greater the energy costs to consumers.

The design of a GHG tax can be straightforward. The application would depend on the size of the emissions source. For large industrial plants, the tax would be applied on the basis of measured emissions. This would ensure that industry has an incentive to capture and store, if that is a cheaper option than fuel switching or energy efficiency. In the case of emissions from smaller sources, such as home furnaces and vehicles, the tax would be applied to the fuel, based on its carbon content.

The tax would rise gradually. Firms and individuals need time to adjust – to make appropriate decisions based on knowledge of what's coming, and taking into account the normal life expectancies of equipment, vehicles, and buildings. A gradually rising tax would therefore provide the correct signal for new product development. Manufacturers and consumers would both understand that the days of free carbon dumping into the atmosphere are over; that a cost would be attached to such action in the future.

Governments are poor micromanagers of economic decisions. They are better at setting overall standards, making regulations, and determining tax rates, providing the context within which individuals and firms make their decisions. One reason economists like carbon taxes is that they reduce emissions cost-effectively. Once the taxes are known, and especially if their levels are spelled out well in advance of imposition, individuals and firms try to cut emissions and figure out where cuts can be made the most cheaply. Over time, as a carbon tax increases, so do the number of activities where cutting emissions becomes cost-effective. Some firms and individuals, even when faced with very high taxes, would keep emitting GHGs from activities for which no ready, carbon-free substitute exists, and from which they derive very high value. This feature enhances the attractiveness of carbon taxes, because they ensure that the lowest-cost reductions are pursued first. Society would gradually achieve its goal of GHG reduction without being forced into either heavy-handed choices or clumsy micromanagement decisions by government about how much GHG each person or industry should be allowed to emit. Landowners could even get credits within the GHG tax system for converting their agricultural land to forest, since trees absorb carbon emissions.

The GHG tax is so potentially powerful that on its own it could achieve the desired emissions reductions by 2050. It is simply a question of setting the tax high enough. But, given the challenge of political acceptability, the tax might have to be combined with a mix of information programs, subsidies, and command-and-control regulations. Information programs, as we have seen, do not work as a policy for reducing GHGs, but some money would have to be spent informing people about the tax: its rationale and mechanics and how to work with it. Especially important would be the information that the tax is *not* fattening government coffers. Subsidies, too, do not work very well as a general policy approach, but they do have limited usefulness when properly targeted. For example, Canada needs research and development into low-emissions and zero-emissions energy technologies. Similarly, studies have shown that subsidies are useful for reducing emissions from low-income housing, where few incentives exist for private landlords. Governments themselves own stocks of low-income (public) housing, and would have to spend money to refit them.

Carbon taxes, properly imposed, would be proportional to the amount of carbon released by fuel combustion. Natural gas, which emits 50 kilograms of CO_2 to produce one gigajoule of energy (an average Canadian car uses one gigajoule to drive 300 kilometres), would be taxed at a lower rate than coal, which emits 90 kilograms of CO_2 to produce one gigajoule.

GENERATING ELECTRICITY THE CARBON TAX WAY

Consider possible scenarios for electricity generation from a carbon tax. Certainly such a tax would trigger reactions from the electricity supply industry, as well as from firms and households that use electricity. If the tax is not too high initially but is forecast to climb gradually over time, that would certainly influence decisions when new investments are made in buildings and equipment, a much cheaper way of transforming society's technologies. Such a tax would gradually increase the cost of generating electricity in a conventional coal-fired plant. Over time, we would expect to see the GHG-emitting generation of electricity from conventional coal-fired plants decline, or even disappear, as future investments, depending on cost and region, are made in some mix of renewables, natural gas, nuclear power, and coal-fired generation with carbon capture and storage. The capture-and-storage option would be cheaper in Alberta and Saskatchewan, where coal deposits are not far from depleted oil reservoirs or saline aquifers. The cost would be higher in Ontario, which imports coal and would have to export CO_2 to more distant suitable storage sites.

Even for Ontario, the cheapest way to reduce emissions rapidly might be converting conventional coal-fired facilities into ones that use carbon capture and storage. Premier McGuinty, asked about this option in January 2007, ruled it out. But other alternatives such as renewables, nuclear power, natural gas generation, and electricity imports from Manitoba and Quebec are also likely to be costly. The incremental cost of developing zero-emissions coal-based generation would be lower because of the existing infrastructure at coal-fired plants – coal ports, coal storage, and power station and transmission grid connections.

In areas of Canada where electricity is mostly generated by coal, the imposition of a carbon tax would mean that the cost of electricity would rise faster than that of natural gas or refined petroleum products. Natural gas would compare very favourably from a price perspective because of its overall lower emissions per unit of energy compared with coal. Power producers in coal-rich regions such as Alberta and Saskatchewan would be motivated to install carbon capture and storage when they make new investments in electricity generation if a price is placed on carbon emissions. A tax would raise electricity prices, but its impact could be small. The final effect on electricity in a province like Alberta might be less than a penny per kilowatt hour, with a rise of less than 1 per cent a year in bills for a typical residential electricity customer – with the overall benefit of shifting toward lower emissions by mid-century.

IN PRAISE OF TAXES

GHG taxes would generate revenue for the government. That revenue, paradoxically perhaps, could become a problem. Politicians do not like – because voters do not like – higher taxes. For more than a decade in Canada, cutting taxes has been all the rage at the federal level and in many provinces. Subsidies are far more politically attractive, although they, too, come from taxes.

Politicians with a modicum of courage can make the case for taxing carbon, quite apart from the major societal benefit of helping to combat climate change by changing patterns of consumption. They could say, Yes, the tax will raise revenue, and here is how we propose to spend it: for health care, education, public transit, investments in low-emitting energy technologies, whatever. Of course the public might reject any kind of additional spending, and economists would worry, not without reason, about the impact of higher taxes on economic growth, incentives, job creation, and so forth.

So politicians have another option, a very good one. They could decide that GHG taxes would be offset by reduced taxes elsewhere, so that the net effect would be revenue-neutral. That is, the government would have the same amount of money from all revenue sources, because some taxes would rise while others would fall. Economists call it "tax shifting." For example, the United Kingdom uses almost all the revenue from its "climate change levy" (a tax on GHG emissions) to lower its national insurance tax. Of course, politicians are aware of the political dangers of the "revenue-neutral" argument. It might be economically sound and factually accurate, but put through the grinder of opposition party attacks and the inanities of the five-second sound bite, all that might emerge is an impression of a tax increase. So honesty demands respect for how carefully politicians would have to put across this case, and how visible the compensating tax reduction would have to be.

Rather than just cutting the GST by another point at a cost of about $5 billion, for example, the government, if it is serious about combating climate change, could impose taxes on carbon worth $5 billion and avoid a gaping hole in government revenues. Or, since Canada derives a larger share of its overall revenues from personal income taxes than most other OECD countries and most economists think that share is too high, revenue from carbon taxes could be used to lower personal income taxes.

Whatever choices are made – and these are only two among many – carbon taxes would put a price on pollution, end the practice of considering the atmosphere a free dump, change behaviour, and bring the "externality" of the environment directly into the hard-headed calculations of firms and households, rather than allowing concerns about the environment to languish as an afterthought.

THE ALBERTA SOLUTION

Ever since the first murmurings about climate change, politicians have insisted that whatever happens in Canada, no one region should be singled out for discriminatory treatment. This was usually a euphemism for protecting Alberta (and Saskatchewan, to a lesser extent), since that province emits far more carbon per capita than any other part of Canada. Nowhere has the reaction to carbon taxes been more resolute, even hysterical, than in Alberta, in part because they are seen (correctly) as directed at the fossil fuel industries but also because of the province's overall emissions record.

But a carbon tax can be designed that does not single out Alberta. The federal government, if it is concerned about this issue, could easily calculate the total amount of GHG tax revenue collected from Alberta citizens and businesses in each year, and then send this amount back to the provincial government or to the individuals and businesses, still retaining the incentive to cut emissions. Alberta politicians and the province's usually compliant media keep insisting that carbon taxes would be devastating to Alberta. That assertion has become a kind of mythology, and like all myths it has become apparently impervious to creative thought or creative analysis. But carbon taxes, designed with rebates in mind, don't have to drain money from Alberta to the rest of the country.

Norway, Denmark, Sweden, and the United Kingdom have implemented and maintained carbon taxes without a substantial political backlash. Norway, an oil and gas exporter like Canada, imposed a GHG tax in 1991 averaging $30 per tonne, and rising as high as $75 per tonne for some sectors. It has since seen economic growth of 40.3 per cent, compared with 23.9 per cent for Canada. Norway's GHG emissions have decreased per capita by 0.2 per cent, whereas Canada's have grown per capita by 6 per cent. Canada could be halfway to its Kyoto target today had it followed Norway's policy of the early 1990s.

REGULATIONS THAT USE THE MARKET

So much for the fear of carbon taxes. They work, as evidence from other countries shows. They can be designed not to raise the overall tax burden and not to burden one region disproportionately. But what other economic measures can help get a grip on rising GHG emissions, in addition to command-and-control regulations and taxes?

Interest has grown in a new category of climate change approaches called market-oriented regulations. These approaches are actually not so new, for they have been effectively used in achieving reductions of air pollutants such as sulphur dioxide, but they are in the early stages of design and implementation for GHG emissions.

Like command-and-control regulations, market-oriented regulations specify a certain outcome, usually in terms of technologies or emissions, but provide a lot of flexibility so that businesses can decide how to respond. They can undertake their own action to reduce GHGs, thereby complying with the regulation, or pay other firms to achieve the reductions. In this sense, these regulations function like a GHG tax without being a tax.

An emissions cap with tradable permits is one form of market-oriented regulation. As we have seen, under this system, a cap on emissions is set for each firm covered by the policy. Permits are issued for emissions up to the level of the cap. The cap declines over time, thereby forcing the firm to reduce its emissions or purchase extra permits from other firms that have reduced emissions below their caps and therefore have permits to sell. This system was first used in a major application through an amendment to the U.S. Clean Air Act in 1990. It produced a dramatic reduction over the next 15 years in acid emissions from electricity generators and major industrial plants. So we know from experience, not just theory, that the system can work.

The Europeans launched a GHG emissions trading system in January 2005, and all large industrial firms in Europe are required to participate. The European Commission sets caps for the total amount of carbon emissions each firm can emit. Firms are allocated permits based on the cap and on emissions in an initial year. Firms remit enough permits to the commission to cover their GHG emissions during the year, with permit trading allowed. The trading price of permits provides a similar incentive to a GHG tax. But whereas a tax guarantees a maximum cost to businesses and consumers, a permit trading system, in which permit prices might fluctuate unpredictably in the market, does not.

Critics have had good sport with the European cap-and-trade system because it has indeed experienced teething problems. Too many permits were initially allocated. Permit prices plummeted in April 2006, losing over two-thirds of their value, as companies realized that there were more permits on the market than emissions.

Through the teething problems, politicians and analysts have learned important lessons from the first two years of the European system. The system will undoubtedly work better in the future. But the early challenges did highlight one difficulty with emissions trading schemes: allocating permits. Ideally, permits should be auctioned by governments to companies to establish a price. But politicians face difficulties trying to force companies to buy permits in the face of claims that they will lose competitive ground to those companies that do not need to buy permits. As a result, politicians usually prefer to grant permits to companies for free.

RENEWABLE PORTFOLIO STANDARD

The challenge of properly allocating permits is avoided under another kind of market-oriented regulation. Instead of capping emissions, the obligation-and-certificate trading system requires companies to use a minimum level of a desirable technology to produce better GHG emissions results. One example of this system is the renewable portfolio standard for electricity generators. Half the U.S. states have renewable portfolio standards, which require electricity generators to demonstrate that they have acquired a specified portion of their power from renewable sources. Electricity companies choose whether to produce the required renewable power themselves or purchase certificates from renewable generators. Allowing this type of flexibility ensures that the lowest-cost renewable resources are pursued first, thereby lowering the cost of achieving the policy goal.

THE CARBON MANAGEMENT STANDARD

Another obligation-and-certificate trading policy is the carbon management standard. The principle is quite simple. Fossil fuel producers and importers would be required to ensure that a growing fraction of the carbon they extract and process does not reach the atmosphere. The carbon management standard would cover all carbon contained in fossil fuels. It could also cover other emissions that contribute to global warming. Projects would be certified through government or third-party audits. Certificates could be marketed through a central emissions exchange.

As with the GHG tax, the obligations would start modestly. No company would be bankrupted by an early draconian imposition of an impossible regulation. The obligation to produce more fuel with zero life-cycle emissions would grow over time. Companies would know this in advance. The regulations would provide long lead times for companies to prepare for the stiffer obligations. The obligation would apply to fossil fuel importers when their product enters Canada. Fossil fuel exporters could receive credits for exported carbon to limit impacts on international competitiveness of the energy sector, and to allow the system to mesh with foreign emissions trading systems.

The carbon management standard would require a growing share of carbon from fossil fuel production to be kept out of the atmosphere. By contrast, a conventional upstream cap-and-trade system would set a limit on the overall amount of carbon that could be sold by fossil fuel producers and importers. Rather than allocating permits to emitters in accordance with the cap, the government would collect certificates from firms that must match their aggregate obligation. Certificates would prove that some carbon from fossil fuel production and use is not reaching the atmosphere. By using an obligation-and-certificate approach rather than the conventional cap-and-permit approach, the government would avoid politically and economically thorny negotiations over initial permit allocation. Revenue generation would be minimized with the carbon management system, thereby increasing the policy's political acceptability.

This carbon management or portfolio approach could also be applied downstream on emitting industries, instead of upstream on the fossil fuel producers. Thus, electricity generators could face a carbon portfolio standard that requires them to have a declining percentage of GHG emissions associated with each kilowatt hour of electricity they generate. Or, producers of vehicle fuels could be required to provide a gradually decreasing carbon intensity of the fuels they sell – a fuel carbon content standard.

VEHICLE EMISSIONS STANDARD

Development of low-emissions and zero-emissions vehicles is critical for generating long-term and deep GHG reductions. Voluntary agreements with manufacturers will not produce these results. What might be needed is a combination of the market-oriented emissions regulations or carbon taxes to influence fossil fuel producers and the vehicle emissions standard to influence a key consumer product like cars. The combination would get results, since the cost of operating emissions-producing vehicles would rise as a result of taxes or regulatory constraint on carbon emissions, and more low- and zero-emissions cars would be available as a result of the vehicle emissions standard. Fossil fuel producers would be converting an increasing share of fossil fuels into zero-emissions fuels – electricity and hydrogen – that are needed in zero-emissions vehicles.

The vehicle emissions standard (VES) is a form of the obligation-and-certificate trading system; it requires vehicle manufacturers and importers to sell by a target date a minimum number of low- and zero-emissions vehicles. The share of the market for these vehicles must rise over time. The long-term goal would be for the production of these vehicles to reach critical thresholds so that costs drop and consumer acceptance accordingly rises, resulting in a market transformation. The VES accelerates development, commercialization, and dissemination of low- and zero-emissions vehicles while letting industry pick the specific technologies. Under the California VES, a per-vehicle penalty of $5,000 is charged to manufacturers that do not sell the required number of vehicles in the low- and zero-emissions categories, but manufacturers can trade among themselves to meet the overall target. Manufacturers receive significant flexibility in meeting the aggregate market outcome. They can quickly bring down the cost of innovative lower-emissions vehicles, and keep costs to consumers low by sharing the incremental costs of clean vehicles over the entire new vehicle fleet.

The VES policy creates a market for low- and zero-emissions secondary sources of energy in the transportation sector such as electricity and hydrogen, as well as biofuels such as ethanol and biodiesel. The VES could work together with the carbon management standard or the GHG tax. The VES ensures production of more vehicles with engine platforms that use fuels other than gasoline and diesel from fossil fuels; the GHG tax would gradually increase the cost of using fossil-based fuels. The VES approach could eventually be extended to all modes of ground and ocean transportation.

California has taken the North American lead with a standard that requires vehicle manufacturers to produce a minimum of zero-emissions and low-emissions vehicles by a given date. Northeastern U.S. states have adopted the California standard, so that states that use the standard now represent almost 20 per cent of the U.S. car vehicle market. Manufacturers can trade zero-emissions credits among themselves. They decide what their fleet mix should be within the standard.

BUILDING BETTER BUILDINGS

The third component of a market-oriented regulation policy package is a low- and zero-emissions standard for buildings. This policy would resemble the vehicle emissions standard, in that it would require a growing number of new buildings over the next 50 years to be built to zero- or low-emissions standards. One example of a zero-emissions building would be a single detached house with a ground-source heat pump that provides space and water heating and cooling. Electricity for this and all other electronic devices in the house would be provided by solar panels. Another example of a zero-emissions house would be one connected to a hydrogen delivery system that meets all its heating, cooling, and electrical needs with a fuel cell. (This would meet the zero test, however, only if the hydrogen is produced in zero-emissions processes such as electrolysis of water, using renewables or nuclear power, or production from fossil fuels that use complete, or almost complete, carbon capture and storage.)

Like the vehicle emissions standard, the building emissions standard would not actually be required once the carbon management standard is in place. But it would foster the development of low- and zero-emissions technologies that work in harmony to provide a market for the electricity and hydrogen that fossil fuel producers would increasingly be motivated to produce and market as their carbon management standard tightens.

Remember that sound policy focuses on emissions rather than energy use. Both the GHG tax and market-oriented regulations would focus therefore on energy-related GHG emissions. Command-and-control regulations are somewhat different. They could be adjusted every decade or so to consolidate market transition caused by the tax, toward significantly lower-emissions buildings and equipment. They could also prohibit equipment that uses energy in standby mode. Here command and control would focus on energy use, because this is one area where energy efficiency gains can be easily realized.

TOOLS THAT WORK

These, then, are the economic policy tools that can work over time to stabilize and reduce GHG emissions. They would induce people and businesses to take rational decisions to do the right things for themselves and the atmosphere, based on market signals and government regulations. They would put a price or regulated constraint on carbon emissions; they would steer people away from the use of carbon-intensive fuels as consumers, and from its production as industries. They would be applied gradually, so that people and businesses have time to make the most cost-effective adjustments and are not hammered by unanticipated cost increases. They rely on basic economic models rather than exhortations and the false signals of subsidies. They take the experience learned from successful emissions reduction programs in the fight against acid rain and other pollutants and apply them to the GHG emissions challenge.

They are, in a sense, old policies of the kind that underpin much of what we consider conventional economics, but they are new in Canada in the battle against GHGs. With novelty will come controversy, as in the political refusal to talk sensibly about carbon taxes. Sooner or later – and we hope for the sake of the country and the atmosphere that it will be sooner – the policies we have described will be the focus of national discussion, so that we can start making real progress instead of kidding ourselves.

CHAPTER EIGHT

The Role of Government and Citizen

How do we measure different policies to know which ones will work? To answer that question, we need a standard of comparison. The best one arguably is a picture of what will happen if we do nothing or continue to apply ineffective policies, as Canada has done for two decades.

No one can predict with pinpoint accuracy a business-as-usual trajectory stretching over many years. But here are some factors that we know will shape our future if we stay on the same track. Canada's population grew 5.4 per cent between 2001 and 2006. It will continue to increase, courtesy of large immigration flows. The Canadian economy will also continue to grow. Canada will continue to exploit its resources of coal, conventional oil, oil sands, natural gas, and unconventional natural gas. In the absence of policies that tax or regulate GHG emissions, the combustion of fossil fuels will remain the favoured option for electricity generation, heating buildings, creating thermal heat for industry, and powering most modes of transportation.

Industry will continue to boast about improvements in energy intensity, but these improvements will be swamped by higher overall output. The outgoing head of the Alberta Utilities Commission, for example, predicted in March 2007 that oil sands production would quadruple by 2025. We know that extracting oil from the sands produces at least twice the GHG emissions per unit of energy as does the extraction of conventional oil. At that rate of expansion, any improvements in intensity will be overwhelmed by the sheer additional volume of emissions. An intensity improvements approach could see reduced emissions per building, system, equipment, and appliance, but these gains would be outpaced by growth in population, economy, and fossil fuel production.

These factors would increase business-as-usual GHG emissions at a rate similar to that of the past decade. The business-as-usual rate, augmented by accelerated oil sands development, means that if we do nothing or continue with the same mix of failed policies, Canada's GHG emissions will climb from more than 800 million tonnes in 2010 to almost 1.4 billion tonnes in 2050. Instead of GHGs being cut by half, as governments keep projecting should happen (and as various international studies insist is required on a worldwide basis), Canada's GHG emissions are headed toward being 50 to 100 per cent *higher* in 2050 than they are today. That projection could be somewhat altered by technological breakthroughs, but if policies do not change, Canada's GHG record, already among the worst in the industrialized world, will continue to place Canada among the world's laggards, to use a gentle word.

Knowing where the business-as-usual approach would take Canada provides a baseline against which we can judge the policies we recommend. When we simulated them in the CIMS model, we set the range of effective policy options outlined – and recommended – in the previous chapter to achieve similar results in emissions reductions. As for financing, all of these major effective policy options – GHG tax, the cap-and-trade system for large final emitters combined with a tax or cap on household emissions, the upstream carbon cap-and-trade system, and the carbon management standard – when combined with standards for vehicle emissions and buildings *will cost about the same to achieve a similar level of* GHG *emissions reductions*. We simulate more and less aggressive versions of these policies. The more aggressive ones are driven by maximizing environmental gains to hit broad targets for Canada consistent with Intergovernmental Panel on Climate Change recommendations; the less aggressive policies minimize costs and take longer to reach those targets.

The next two figures present projected business-as-usual GHG emissions in Canada up to 2050 alongside reductions of GHG emissions under the more and less aggressive policy scenarios we developed. The GHG reductions are depicted as wedges that adjust over time as capital stock turnover (power plants, factories, homes, vehicles) converts the economy to low- and zero-emissions technologies and activities. The wedges depict the four major categories of actions: land use changes (in agriculture and other sectors), carbon capture and storage, fuel switching, and energy efficiency.

Those who still claim that Canada can reach its Kyoto target of a 6 per cent reduction from 1990 levels by 2008–12 should study these figures carefully. Note that the emissions projection for the aggressive policy scenario shows that not until after 2025 – *even with the immediate implementation of these strong policies* – would Canada reach its Kyoto commitment, or about 560 million tonnes.

The good news is that aggressive policies do produce dramatic longer-term results. Follow the emissions line and you will see that the aggressive policy portfolio would bring by 2050 a 50 per cent reduction in emissions below 2010 levels, and a 70 per cent reduction over the business-as-usual scenario. The IPCC recommends that by the end of the century, global GHG emissions must fall by at least 80 per cent from today's levels. It adds that

Canadian GHG Emissions – Aggressive Policy Scenario

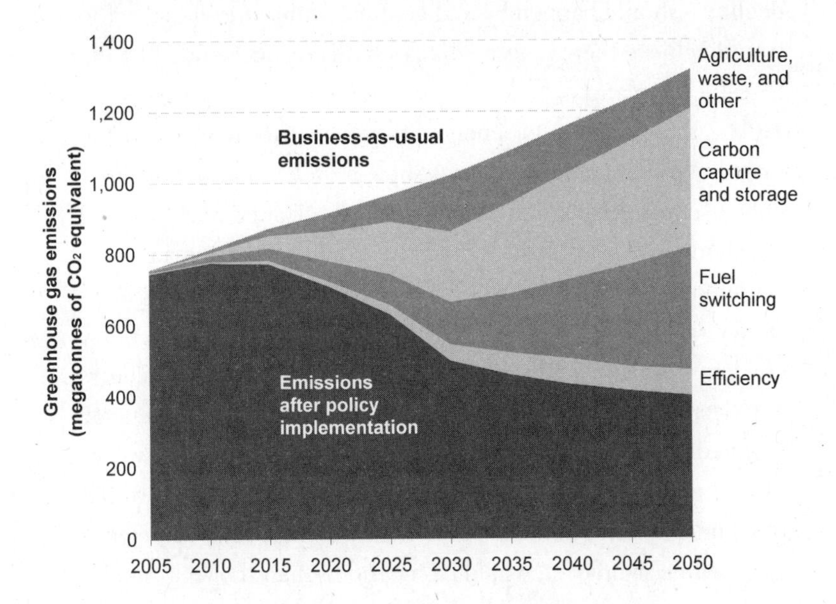

Canadian GHG Emissions – Less Aggressive Policy Scenario

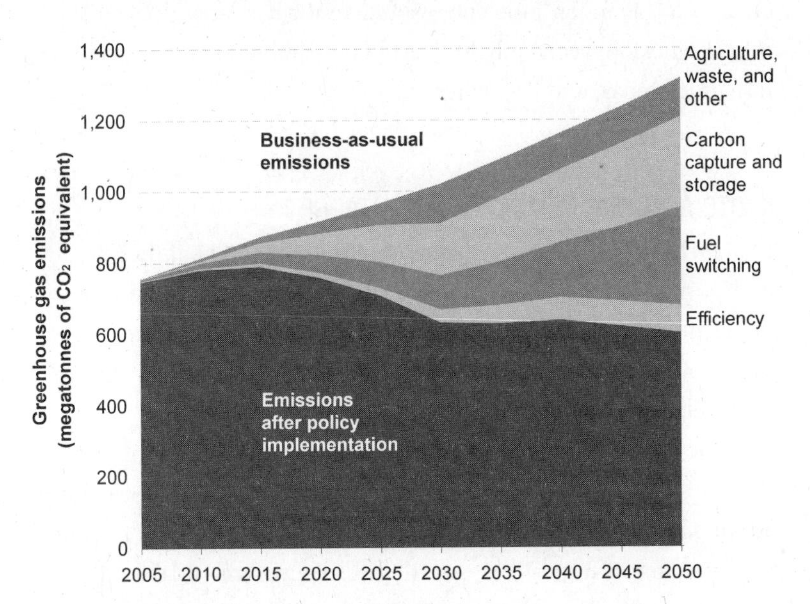

Source: Calculations by Mark Jaccard and Nic Rivers.

a considerable amount of this reduction – say, a 50 per cent decline – should happen by mid-century. Our aggressive scenario would achieve the 50 per cent goal that is recommended by the IPCC for the planet.

The strange news, for energy efficiency advocates, is that the CIMS policy simulation does not show a great contribution from energy efficiency. Actions that switch to cleaner fuels, control emissions, and even change land use would make a greater contribution to emissions reduction. One reason is that the energy efficiency wedge shows the net effect of improvements in energy efficiency in many end-use technologies in industry and buildings and decreases in energy efficiency resulting from the extra energy required to capture and store carbon, and to develop renewables like biomass. The losses in efficiency from these activities would offset much of the efficiency gains. In any event, end-use efficiency gains, while significant, would be much less than shown in studies that look only at technologies but not costs and consumer preferences. Not as much efficiency would result if energy prices rise over the next 40 years only to a level comparable to energy prices in Europe today, and no higher.

KYOTO AS A WAR MEMORIAL, NOT A TARGET

Politics being politics, short electoral cycles being what they are, and purveyors of quick-fix solutions and five-second sound bites being all around us, the Canadian debate has been about meeting the Kyoto target. This focus, as we have tried to show, creates a false debate that emphasizes the undoable and the unwise rather than the doable and the strategic. The debate is about symbols and rhetoric, rather than serious policy alternatives. Painful as it is to admit, Canada has blown Kyoto. The country negotiated a target that was almost impossible, and then the government refused to apply the only policies with even a remote chance of meeting it, discarding the advice and warnings of scientists, economists and senior officials to that effect. For many years, Canadians were misled about what might work in combating climate change. They were told about targets but not about the path to achieving them. Politicians, then and now, defend themselves by saying Canadians were not ready for the truth; critics can reply that they never tried honesty, so the public can hardly be blamed for preferring bromides and easy non-solutions.

MORE AND LESS AGGRESSIVE POLICY CHOICES

Policy choices do involve some trade-offs between environmental risks and economic costs. Canadians cannot make those choices intelligently unless they are given the facts. With the CIMS projections, and knowing from experience which policies do not work, Canadians might now understand the real choices. Even now our politicians are not levelling with Canadians by explaining frankly that only regulatory and pricing policies can lower GHGs. In these projections, the trade-offs are reflected in the alternatives of aggressive and less aggressive policies. Canadians can make their choices.

It's too bad, even a tragedy, that this kind of analysis and public information was not provided to Canadians before Canada signed the Kyoto Protocol in 1997. The course of our history and policy would likely have been very different. Maybe Canadians, as they thought about alternatives, would have recoiled in horror and preferred the paths of least resistance and effectiveness; maybe, just maybe, they could have been persuaded that the tough policies were indeed the preferable ones because they would produce effective measures to cut GHG emissions.

In any event, as can be seen from the previous pair of figures, the less aggressive policy of taxes and market-oriented regulations would produce a greater reduction than a business-as-usual approach, but it would not get Canada to the 50 per cent GHG reduction target by 2050. The less aggressive scenario has more modest market-oriented regulations, but its principal difference from the aggressive approach is in the cost it assigns to CO_2 emissions, as shown in the next table. If the policy is a tax, these would be the prices per tonne of CO_2, with equivalent charges for other GHG emissions like methane. If the policy is emissions cap and trade, these would be the prices of permits in the permit trading market. In the more aggressive scenario, to achieve the 50 per cent reduction by 2050 would require a carbon tax that rises to $180 per tonne by 2030. This level would exceed what most experts say is needed to produce market dominance for low- and zero-emissions technologies. Canada would therefore be bearing additional costs to replace capital stock before replacement was otherwise necessary, because of the tight time frame of the emissions target. Although it is more than 40 years away, the 2050 target does not leave much time to create a dominant market position for low- and zero-emissions technologies, given the rate of stock turnover in the economy.

Permit Price or Tax per Tonne of Carbon Dioxide under Cap and Trade

	2010	2015	2020	2025	2030	2035	2040	2045	2050
Less aggressive policy	$15	$20	$60	$100	$120	$120	$120	$120	$120
More aggressive policy	$15	$20	$60	$120	$180	$180	$180	$180	$180

Emissions would still fall significantly in the less aggressive scenario, but not to the same extent as in the aggressive one. At a carbon tax level of $120 per tonne, the costs to the Canadian economy would be lower. There would be virtually no forced premature retirement or replacement of buildings and equipment as the economy moves toward a low-emissions future. The pace of change would match the natural rate of turnover of capital stock.

In considering the choices before them, Canadians should not fret unduly today about the specific target for 2050. Under both the aggressive and less aggressive scenarios, our recommended policies would be virtually the same for the next decade, be they market-oriented regulations or GHG taxes. In both scenarios, the GHG tax would be $20 in 2015 and $60 in 2020. Of course, these tax levels would not be cast in stone. They could be adjusted according to whether emissions are falling more or less rapidly than anticipated, and whether the costs are proving to be more or less than forecast. They should also take account of policy choices in other countries. As all this new information is absorbed over time, adjustments can be made to the long-term schedule for GHG taxes or market-oriented regulations. What counts more than fixing policies in stone is getting the right policies at the beginning, laying out a clear framework within which individuals and business can begin to plan, and adjusting the framework, timelines, and details as experience dictates.

EFFICIENCY ISN'T EVERYTHING

Advocates of energy efficiency, especially environmentalists and some politicians, will not like the lessons to be drawn from these simulations. They show that energy efficiency, the Holy Grail of so many environmentalists, would not play the dominant role in GHG reduction that so many of them believe it will and must. They don't like the independent studies that have shown energy efficiency often costs more than advocates contend, in part because of the extra risk inherent in long-payback technologies and in part because efficient technologies are rarely perfect substitutes for the technologies with which they purport to compete. Information programs advocating energy conservation and efficient devices cannot compete with the deluge of advertising about products and services that urge consumers and businesses to acquire the latest energy-using innovation. And it is difficult to prevent free riders from capturing half or more of the subsidies designed to encourage people to acquire more efficient devices.

Energy efficiency therefore has its place, and one that many citizens try to pursue in their daily lives. But its place overall is not the predominant one its enthusiasts claim. Other choices would become more important than energy efficiency once taxes or market-oriented regulations are put in place, people's preferences are considered, and a full slate of technological options are presented. Take, for example, the effect of a rising price for carbon-based fuels. Initially, the rising price would induce some consumers to purchase smaller, more efficient cars. Some people might even opt to take public transit, because of rising fuel prices rather than because of the Harper government's tax credit for transit passes. But many other vehicle owners would continue to prefer vans, pickup trucks, or SUVs. Increasingly, as the carbon tax rises, they would adopt variations of these vehicles that are powered by biofuels or electricity with a small amount of fuel (a plug-in hybrid, for example), or fully electric vehicles, or perhaps hydrogen fuel cells in a decade or two (with the hydrogen produced in processes that do not cause emissions). The most likely scenario is that the response to the rising cost of carbon-based fuels over the long term will be fuel switching and emissions control rather than fuel efficiency. (Remember that the rising cost of carbon-based fuels should be offset through tax reductions elsewhere so that the net tax effect is neutral.) This example from the world of the family car typifies decisions that will be made throughout the economy, from the largest industrial plant or shopping centre to the choice of a backyard patio heater.

Carbon capture and storage is thus likely to play a large role in some regions. In Alberta and Saskatchewan, major fossil-fuel-producing provinces, coal with carbon capture and storage would be among the cheapest options for generating electricity, especially when captured carbon could be used to enhance oil and gas recovery – a system already working in Weyburn. A rising carbon tax and/or ever tighter market-oriented regulations would motivate industries such as oil refining to capture carbon. We understand very well, based on experience, that they will not move aggressively if left to their own devices, nor will improvements in energy intensity get us remotely near the reductions required from fossil-fuel-producing industries and other large emitters.

HOW WOULD DIFFERENT SECTORS RESPOND?

We have used the CIMS model to simulate how various sectors of the Canadian economy would respond over the next four decades to aggressive or less aggressive taxes and market-oriented regulations.

GHG Emissions in Canadian Economic Sectors (megatonnes of CO_2 equivalent)

	2010	2050 Business as usual	2050 More aggressive policy	2050 Less aggressive policy
Electricity Generation	127	178	23	35
Oil and Gas Production	176	325	117	203
Energy-intensive Industry	112	194	59	93
Non Energy-intensive Industry	23	66	22	33
Residential	41	19	7	9
Transportation	193	272	95	136
Services	42	102	33	49
Other	100	156	43	43
Total	**813**	**1,313**	**400**	**601**

Source: Calculations by Mark Jaccard and Nic Rivers.

The electricity sector would prove highly responsive to aggressive policies. It could produce reductions of 90 per cent by 2050, compared with 2010 levels. These would come through switching the fuel for generation to nuclear or renewables, or through carbon capture and storage. Choosing among these options would largely be a political decision. Given the difficult politics of nuclear energy, the limited number of additional available sites for hydro, and a realistic assessment of how much energy the renewables could actually produce (to say nothing of the not-in-my-back-yard opposition to wind power sites), carbon capture and storage would do well in some regions.

In industries other than electricity generation, reductions would come from carbon capture and storage; energy efficiency, and switching to low carbon fuels. Energy efficiency advocates and most environmentalists argue that energy efficiency will play a dominant role in GHG reduction, with some even suggesting we can reduce energy use by 50 per cent through efficiency gains. Our simulation of effective policies, by contrast, suggests that energy efficiency's contribution might be much less once offsetting reductions in efficiency in the energy supply sector have been incorporated. A greater role would be played by fuel switching and pollution control.

Oil and gas production would also be quite responsive to aggressive GHG policies over the long term, with emissions falling significantly. Major industries would be both more energy-efficient and less polluting. Transportation and non-energy-intensive industries would reduce emissions relatively less, because the cost of emissions reduction in these sectors would be higher. But by 2050, with aggressive policies, new vehicles would be almost entirely zero-emissions, and buildings would be constructed to considerably higher standards of energy efficiency and would consume less fossil fuel directly.

POLICIES THAT WILL WORK

These are the projected results of aggressive and less aggressive policies to reduce greenhouse gases in Canada by using regulations and pricing policies. The anticipated results are realistic ones, not the promised but unrealized hopes of the past or the unsatisfactory ones Canadians can expect if they count on the Harper government's approach, a reworking of previous Liberal policies, or Stéphane Dion's newest policies.

These options would work: an economy-wide GHG tax; the carbon management standard; the upstream cap-and-trade system; and an emissions cap-and-trade system for large final emitters, combined with either a carbon tax for smaller stationary and mobile emitters (households, commercial and institutional buildings, transportation equipment) or a decentralized cap-and-trade system. Canadians and their politicians can choose which specific policies to apply if they are serious about Canada doing its part to combat climate change.

BUT HOW WILL THE NEW POLICIES AFFECT ME?

Taken together, these policies would reduce GHG emissions. But Canadians need to know in more detail, while mulling over the options, what impacts they could expect on various aspects of their national and personal lives. What would these policies do to the economy? What would happen to individual taxpayers? To regions, especially those such as Alberta that are large GHG emitters? To the costs of electricity, driving, heating? We can sketch out, based on simulations, the answers to these reasonable questions. We cannot be precise, because the answers depend on the choices governments, consumers, and producers make, but we can certainly offer solid projections. For example, we know that people would still use vehicles, some of which would be as large as those on the road today. We also know that vehicles would be more efficient and be zero-emissions, or close to it. But we cannot predict which share of vehicles on the road would be plug-in hybrids, battery electric, or powered by biofuels. Maybe hydrogen fuel cells will be the breakthrough technology of the future. What we can say with reasonable assurance is that it won't matter which technology pulls ahead as long as they all lead us to low- or zero-emissions vehicles.

Overall, as we have shown, by 2050 the more aggressive policy scenario would decrease Canadian GHG emissions by 50 per cent from their levels of 2010. Such a reduction would reflect a huge number of significant technological changes, moving us away from an economy that is overwhelmingly driven by the combustion of fossil fuels. These changes would create business opportunities. Some regions would benefit from the changes, sometimes dramatically. But an aggressive approach would also entail costs. Some of these would be concentrated in specific parts of the country. These could be mitigated, if governments choose, by designing policies to share burdens equitably across the country. If this were to become the overriding consideration of selling the policy, then straightforward ways exist to design the policy to minimize the impact on any one region.

GOOD NEWS FOR ALBERTANS

Consider the GHG tax policy. Such a tax, even the mere hint of the prospect of such a tax, sets off howls of outrage in fossil-fuel-rich Alberta, with echoes in Saskatchewan. If people would calm down, rather than ranting or conjuring up conspiracies, they would see that a policy could be designed to deal with their fears.

Either level of government could apply a GHG tax. If Ottawa did so, all of the tax revenue could go into a fund that would be redistributed to each province according to its tax contribution. If a GHG tax raised $300 million in Alberta in a given year, then $300 million would be returned to Alberta. The same could apply to households. Each household would receive a cheque for a similar amount, administratively the simplest approach. Or households could be reimbursed more precisely, since utilities have household-by-household records of fuel consumption for billing purposes. GHG taxes paid by fossil-fuel-fired electricity generators could be reimbursed to each customer according to the share of residential electricity it consumed before the tax was launched. Electricity prices would rise to reflect the higher cost to electricity generators resulting from investments to reduce GHG emissions and payment of the GHG tax, but customers would be compensated for GHG tax payments.

The next figure shows the effect, in Alberta, of an aggressive policy on the price of electricity, but also the much lower increase in electricity bills after the tax reimbursement. Under an aggressive scenario, the cost of electricity to Alberta citizens would rise – but by less than 1 per cent a year.

Electricity Cost to a Typical Alberta Household

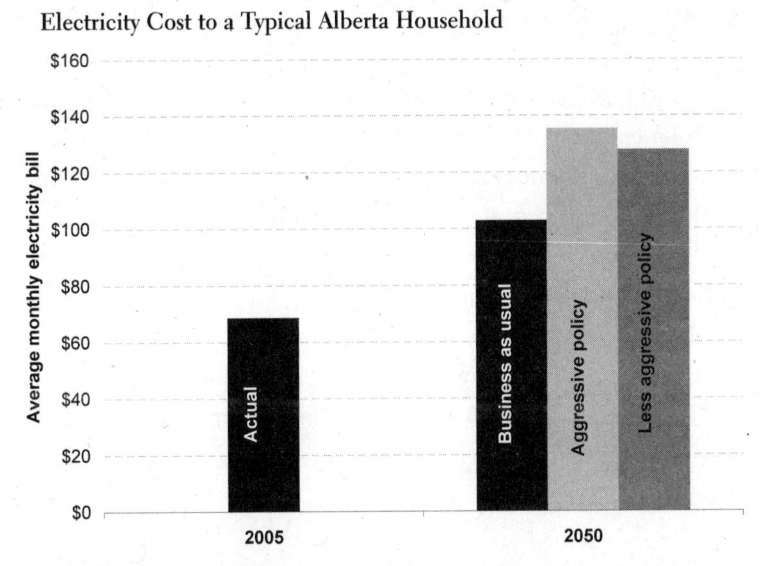

Source: Calculations by Mark Jaccard and Nic Rivers.

Fossil fuels would continue to be used in Alberta, and might still even play a role in Ontario. Canadians would actually be using a lot more electricity for vehicles and heating, despite predictions that energy efficiency gains would lower demand for electricity. The next figure shows the likely sources of electricity generation for three large provinces. Fossil fuels would continue to dominate in Alberta, and hydro power, with its new partner wind, would be the major source in Quebec. Ontario, which has the greatest uncertainty, would continue to shift toward renewables, some of it from domestic sources and some from hydroelectricity imports from Manitoba and Quebec. Of great uncertainty is the relative contribution of coal and natural gas. Will the McGuinty policy of phasing out coal, already delayed by at least seven years, be extended even further, or perhaps indefinitely if carbon capture and storage proves viable?

Albertans have been conditioned by industry spokesmen, their politicians, and some of their clangorous media to believe that any economic policies such as a GHG tax would cripple their economy, devastate their fossil fuel industries, lead to a flight of capital, and represent "Central Canada's" last revenge on the province's prosperity. It will come as a surprise to those so conditioned that the aggressive policy scenario would not have a significant effect on the total output of the energy sector in Alberta. Exploitation of the oil sands would continue to grow – but more carbon would be captured and stored at production facilities, in part to reduce emissions and in part because output would have shifted from carbon-containing petroleum and refined petroleum products to carbon-free electricity and hydrogen, which would then be shipped to markets via transmission wires and pipelines.

We have simulated in CIMS the effect on Alberta's gross domestic product from our aggressive GHG emissions reduction policies, assuming that the tax money raised stays in Alberta. Alberta's GDP

Sources of Electricity Generation, 2050

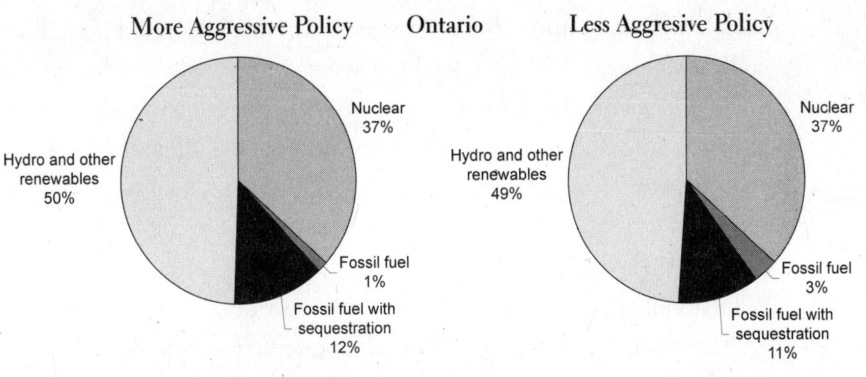

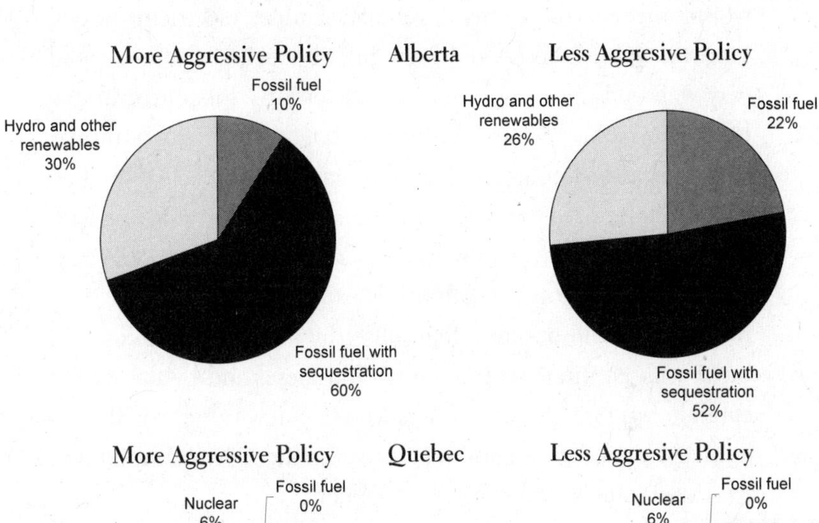

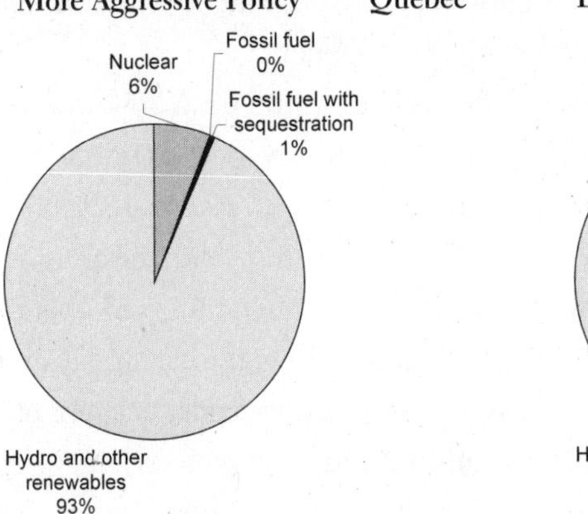

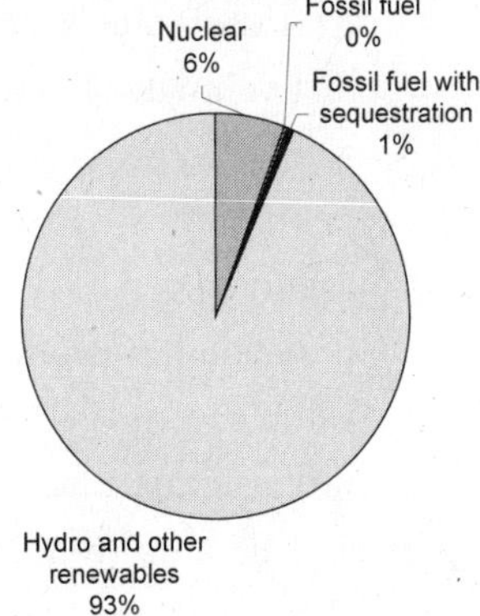

Source: Calculations by Mark Jaccard and Nic Rivers.

would be almost the same by mid-century under a business-as-usual scenario as under the aggressive GHG emissions reduction policies. There would be a slight slowing of oil sands development, and slightly higher electricity and fuel costs affecting other sectors, but overall economic activity would be little changed.

WHAT ABOUT VEHICLES?

What about vehicles? Just as electricity prices would face upward pressure with a GHG tax or with market-oriented regulations, so would the cost of using a vehicle. Unlike electricity prices, however, vehicle costs would not differ by region and they could very well end up being lower by mid-century. Take the example of an automobile owner who drives about 20,000 kilometres per year. In the first couple of decades, gasoline- and diesel-fuelled cars would remain the norm, and their fuel costs would rise with the GHG tax or equivalent market-oriented regulation. Some consumers would choose smaller, less expensive cars, but this cost reduction would be more than offset by the rising fuel costs for those who continue with gasoline and diesel and by the higher capital costs for those consumers who buy new vehicles with low- and zero-emissions engine platforms, including hybrids, plug-in hybrids, biofuels, and hydrogen fuel cells.

Eventually, however, the policies would push down the capital costs and fuel costs of these alternative vehicles, and the average annual costs could actually fall below the costs in a business-as-usual scenario. While there is still much uncertainty in this projection, it is unlikely that the costs of owning and operating a vehicle would be substantially higher than they are today. In large urban centres, however, the costs of using roads (because of tolls) and parking would probably be much higher, as recent trends suggest. Causes of concern about automobiles would evolve, from their emissions to their effect on urban livability. Policies to address congestion and other ills in cities would be likely to have a greater effect on vehicle use patterns than concerns about climate change.

People will still use vehicles in decades to come. We're not going to walk long distances through the snow. Rural people are not going to take public transit to visit their neighbours. Suburbanites heading from one suburb to another are unlikely to take the subway, assuming one exists. People will want the convenience of cars. We are confident, however, that the right pricing and regulatory policies will produce low- or zero-emitting vehicles, so that cars could even be as large as many are today.

The continued pressure on energy consumption, even with charges or regulatory constraints on emissions, reinforces the need to produce electricity from cleaner sources and to capture and store carbon emissions from fossil fuels. Biofuels such as ethanol from grain and especially from cellulosic materials such as straw, grasses, and wood will be more important by 2050. Trucks, trains, and boats will also be shifting away from fuels produced by petroleum. Trucks, which are now heavy emitters, will continue to play a role of cardinal importance in point-to-point deliveries of goods, but their emissions over time will fall as zero-emissions technologies and biofuels increase. Only air transport might still be relying primarily on jet fuel refined from fossil fuels, although even here it is possible that biofuels will make great inroads.

THE PUBLIC TRANSIT DILEMMA

Environmentalists will be disappointed, we predict, in the very modest shift toward public transit. They have been among public transit's biggest boosters, seeing in it an important part of the answer to GHG reduction. There are compelling reasons why more public transit is a good idea, since it makes for more livable cities, cuts down on traffic jams, and eases stress for some of the urban and suburban population. It can also be hugely cost-effective for those who want point-to-point transportation. About 9 per cent of commuters in major cities in 2000 took public transit, but our simulation suggests that only about 14 per cent would do so in 2050. That's a gain, all right, but the vast majority of people would continue to use cars in urban and especially suburban areas. Government decisions about the pace of expanding public transit and densification of urban areas would affect the share of total commuters using public transit – but only at the margin. The bulk of emissions reductions must still come through changing the fuel mix and the kind of vehicles we drive. And, of course, public transit vehicles like buses must shift to lower-emissions fuels and technologies, too.

As Statistics Canada showed in its analysis of the 2006 census, growth in suburban areas eclipsed that of central cities between 2001 and 2006. Suburban areas during that period grew by about 900,000 people, central cities by 500,000 people. Despite efforts to increase central-city density, some of which have been quite successful, most of the growth in urban centres occurred in the rings around the central core. Also growing faster than central cities were satellite cities, part of what urban planners call exurbia, places such as Barrie, Whitby, and Brampton near Toronto; Okotoks south of Calgary; Vaudreuil and Blainville near Montreal; and Surrey and Langley east of Vancouver.

Commuter trains can – and should – absorb some of the movement between exurbia/suburbia and downtown, but a lot of people aren't going point-to-point from exurbia/suburbia to central-city locations. They are going all around suburban and exurban areas, so for them public transit is of limited use. Efforts to increase density, create self-contained suburban areas with bicycle paths and walkways, and build point-to-point public transit are all happening in Canadian cities, and we applaud them as environmentally and socially valuable. Demographic trends, however, are pushing people toward suburbia and beyond in search of more space, cheaper housing, the lifestyles they prefer, and the automobile travel they need and often prefer. So although better public transit is important, and will if properly conceived and financed attract a larger share of commuters, the car will dominate travel in Canada for the foreseeable future. This reality – and it will be a reality no matter what hard-line environmentalists might say – means we must adopt the aggressive policy mix we recommend that severely limits emissions. It is the policy of reason and practicality as against those some environmentalists prefer, which are based on morality and romanticism.

INDUSTRIAL COMPLEXITY

It's hard to develop any picture of industry four decades hence and be sure what will happen. The health of industrial sectors is obviously important to the companies in those sectors and to the overall economy. Each sector's vulnerability to changes will be different. For some, energy costs are small enough and the sector's competitive position strong enough that GHG taxes or market-oriented regulations would have only a marginal effect. In other instances, however, industrial sectors (cement, for example) could be quite vulnerable. Just how vulnerable, if at all, depends upon what policies are adopted by our principal trading partners.

For many years, it was assumed by many, and asserted by those who wanted little or no action in Canada, that the United States would not be moving on climate change. The country did not ratify the Kyoto Protocol. President George W. Bush's policies have been more about finding additional oil supplies than about curbing energy use. His only aggressive policies were putting money into technological research and pouring subsidies into corn ethanol, to the delight of Midwestern farmers and politicians from farm states. Since Canada does so much cross-border trade, the resistance of the United States to action on climate change offered a rationale for limited Canadian efforts, the argument being that any additional costs for GHG reduction imposed on Canadian companies would harm their competitive position within North America.

That argument is now passé. The United States is likely to be moving decisively to reduce GHGs in the near future. The Democratic Party is committed to serious national policies. So, increasingly, are some leading Republicans. At the state level, in addition to California under Governor Arnold Schwarzenegger and the Northeastern states, many others are starting to move. Silicon Valley capitalists, the incubators of the high-tech boom, are switching portfolio money and venture funds into the green economy. The Silicon Valley entrepreneurs sparked one broad revolution; they are determined to be in on the ground floor of another. Huge pension funds, banks, and investment houses such as Goldman Sachs are allocating billions of dollars for investments in green technologies. In short, American capital and many U.S. politicians are leaving the climate change deniers in the dustbin of history, dismissing the legacy of George W. Bush as years of lost opportunity, and pushing their decentralized but dynamic country toward serious action.

As always, action does not come swiftly in a country whose political system was designed by its founding fathers to prevent speed and require deliberation, but when that system eventually settles on what is the proper course of action, it works with remarkable speed to find consensus. Canadians had better watch out, because Americans are aiming to grab a big slice of the economic action that will come from new technologies and other innovations of the green revolution.

No one can be certain where and how the United States will move, but move it will. When it does, that will be the end of the argument by those who have insisted Canada should not do anything serious. Instead, when the United States does move, it could leave Canadians quite embarrassed, since Canada would be trailing the United States on climate change efforts, a blow to Canada's habitual sense of moral superiority toward its American neighbour. Obviously, too, U.S. action would lessen the adverse impact of GHG reduction policies on Canadian industrial sectors.

If, somehow, the United States does not move, or chooses to string out its serious actions for some years, Canada could protect certain sectors that might be disadvantaged by domestic measures. (Oil and gas production, even of the higher-cost oil from tar sands, is unlikely to be affected as long as the international price of oil remains above $40 a barrel on average through to mid-century, despite the lamentations and dire predictions of the oil patch.)

GOOD NEWS

The good news is that because what we are proposing relies on tougher measures down the road, rather than on immediate severe regulations, Canada can launch policies now – as it must, to hit 2050 targets – but adjust them over the next decade, depending on policies in the United States and elsewhere. If particular sectors find themselves under the competitive gun because trading partners, especially the United States, are not taking serious action, Canadian policies can be adjusted. Also, in the unlikely event that the concerns of international climate change scientists diminish, Canada could water down some of its policies. With those caveats, we project that the gross domestic product – the overall growth of the economy – under either the aggressive or the less aggressive policy scenario will be roughly the same as if we continue business as usual. The scary talk about economic catastrophe caused by serious policies to combat climate change is misplaced.

We expect other outcomes as a result of aggressive policies. For example, the share of agricultural land dedicated to what we might call traditional food production would fall to 77 per cent, instead of 99 per cent in a business-as-usual scenario. The change would occur for two reasons. Perhaps 15 per cent of agricultural land would be used for producing primary material for biofuels, and about 8 per cent would be reforested, because forests retain a greater amount of carbon than fields. In due course, trees will become an increasingly important source of cellulosic ethanol.

Buildings, especially new ones, would be constructed differently, with new power sources, so that perhaps 60 per cent of buildings by 2050 would be low- or zero-emissions. This would happen under an aggressive policy mix because of the rising costs of emitting GHGs and the gradually tightening energy efficiency and emissions standards for buildings and appliances.

We are certain that there would be significant, even in some instances radical, changes in equipment, fuel mix, buildings, and transport resulting from aggressive policies such as those we have outlined. But the effect on economic growth would be minimal to nil, and the effect on reducing greenhouse gases would be dramatic. The key is to combat emissions by using regulation and pricing policies, introducing them as soon as possible with quite modest application in the first instance, then gradually tightening them over the forthcoming decades – with a schedule that is widely understood to be rising significantly.

The simulations we have presented show what the effects would likely be for people, businesses, and the general economy – and for GHG emissions. Our proposed policies would certainly beat anything Canadian governments have tried. They are controversial at first blush, until the failed alternatives are considered – at which point their necessity becomes obvious. They are being tried and are working elsewhere, especially in Europe. As in so many other areas of public policy, they remind us that today's heresies are tomorrow's conventional wisdoms.

Conclusion

Canadians fancy their country to be a moral superpower. Nothing better captures this inflated sense of national self than the ChaptersIndigo bookstore chain's slogan: "The World Needs More Canada." As if from the Security Council of the United Nations to the hill tribes of Southeast Asia, leaders and people of every race, creed, and colour are waiting, even begging, for more Canada. A national conceit suggests they are, or should be; whereas in fact, of course, they are not.

This sturdy Canadian mythology was juxtaposed for many decades against the reality that the world, whether it needed more of Canada or not, had been seeing less and less of it. Until the last few years, Canada's foreign aid budget declined for a long time, defence capabilities atrophied, and diplomatic resources ran down. Today, we still have a far higher number of aid officers and diplomats in Canada, as a share of total personnel, than almost any major industrialized country does in its own territory, and for a simple reason. It costs less to station people in Ottawa than abroad. We've been too cheap to do anything about this state of affairs. Canada's foreign aid budget is inching up, but after years of decline. As for the military, equipment purchases are now being belatedly made and recruitment efforts are intensifying. But Canada can still barely manage to undertake one serious mission with about 2,500 personnel, such as in Afghanistan, because of the paucity of soldiers, *matériel*, and money.

This decades-long pattern of being a moral superpower in our own minds, while failing to deliver resources and results, is at one with Canada's climate change record. We made grand statements. We beat our chests. We gave ourselves Herculean targets. We made sure that the world heard about them. We made promises – and then we did not keep them. We thought maybe we could keep them, but we fooled ourselves, part of a pattern of fooling ourselves about the world needing more Canada while giving less of ourselves to it.

When it comes to greenhouse gas emissions reductions, however, the world does need more Canada. Not that our efforts alone will turn back the tide of global warming, but because we contribute to climate change as one of the world's largest per capita emitters. Also, if countries such as Canada – which still does enjoy an enviable international reputation and possesses excellent human and technical capabilities – falter, then it is hard to imagine other less favoured countries being inclined to act.

Many things went wrong with Canada's attempts to grapple with climate change, as this book has illustrated. Certainly our political culture and federal system frustrated action. From the beginning of climate change discussions in Canada, there was a lack of political honesty about what would have been realistic emissions reduction targets for Canada, and what serious measures would have been required to meet them. Alas, the lack of honesty remains in too many quarters.

Support for climate change action has been conflated in the public's mind with support for the Kyoto Protocol, according to polling evidence. The conflation is easily understood, if regrettable, since so much shorthand debate has swirled around "Kyoto," "Kyoto targets," and the "Kyoto process." Too many politicians, being hypersensitive to polls, have therefore continued to assert that Canada can somehow – after doing nothing effective for a decade – suddenly, miraculously, cut GHG emissions by at least 35 per cent in the next three to four years in order to meet our ill-advised initial Kyoto commitment. Those who make this case ought to know better. If they do not, then their words are a terrible indictment of their lack of understanding. We cannot buy our way out of past failures by quickly throwing billions of dollars at international emissions permits or investing in hasty, ill-conceived projects in developing nations. This makes no sense economically, and it would do nothing to get Canada itself moving in the right policy direction. This approach would create an administrative swamp. And it is not realistic politically.

Canada's Kyoto failure does not mean we should abandon support for international efforts to reduce GHG emissions. Of course, Kyoto had its flaws, but any first-time effort to achieve voluntary agreement by countries around the world to accept GHG reduction commitments was bound to be less than perfect. The negotiators at Kyoto were trying to secure an initial agreement among developed countries that would eventually widen to include all major emitters in the world. This ambition was always a daunting one, made more difficult still by the later refusal of the United States to follow through. The post-Kyoto efforts will go on; indeed, preliminary negotiations are now under way for another agreement, and Canada is part of them.

What is now needed, rather than chasing the chimera of Canada's first Kyoto commitment, is a sustained focus on what we need to do to reduce our GHG emissions by half, or more, starting now and heading into the next decade, and continuing for many decades thereafter. We could start by stating the obvious to the world community, even if domestic environmentalists howl: we messed up terribly. The world knows this already. Then, rather than grovelling in self-abasement, we can show a new resolve to make up for our poor record by adopting better policies for the future. If we truly decide to match deeds with image and resolve to be a genuine superpower in the fight against climate change, then the shame of our abandoned first commitment to Kyoto will be forgotten. Our challenge is to learn from our own past failures, and so turn these lessons into future successes.

As talks continue about a post-2012 international climate change treaty, Canada starts from far behind in setting our next commitments. That underscores why we must now get serious, but we should not get caught up in a futile effort to atone for past negligence within an unrealistically short time frame. The intent of the original Kyoto Protocol was to engage the rich countries; it was unsuccessful in securing reduction commitments from developing countries, including huge ones such as China and India. Bringing them along will be an important objective to pursue through a blend of suasion, example, technology transfer, trade threats, and foreign aid.

We can learn from our recent past what not to do: how not to negotiate internationally, how not to make commitments we cannot keep to others and ourselves, and how not to pursue policies bound to fail. We ought to know these things by now, although watching the political parties roll out their climate change ideas has not been edifying or reassuring.

Canada made a commitment at Kyoto that was the most difficult and expensive in the world. It took no account of our economic growth, population growth, cold temperatures, vast distances, and fossil fuel production; it contravened our federal system, because Ottawa broke a fragile federal-provincial consensus on the eve of the negotiations. So the first lesson must be to remember these factors – and therefore undertake international commitments and implement serious domestic policies that take these distinctly Canadian factors into account, because no other country has all these characteristics.

Then we launched an interminable series of internal negotiations among federal and provincial governments and "stakeholders," from which emerged a truth so obviously but politically terrifying that it was never acted upon: financial penalties and regulatory constraints had to be part of an immediate effort to reverse the upward thrust of emissions, without which the country's Kyoto target had no chance of being met. This was the truth that emerged behind closed doors, but it remained there, so frightened were those who discovered it of the likely political reaction. So the second lesson from past failure is to relearn what was conveniently forgotten: the only policies that will work place a price on GHGs or regulate their emission and therefore compel understanding that the atmosphere will no longer be a free trash bin into which GHGs can be dumped.

We witnessed policy prescriptions piled atop green reports that led to information and subsidy programs that did nothing to lower GHG emissions, for excellent and predictable reasons, confirmed by evidence around the world: these approaches do not work. The third lesson from unhappy experience is to stop digging: to realize that although giving the public information about climate change is useful, even important, and that subsidies for research and specific sectors such as low-income housing make sense, no serious climate change policy can be built on them. Information and subsidies are as clay to housing: useless as foundation, useful as exterior cladding.

Throughout the drama, we listened to politicians who purported to lead but did not. The failed leadership began with Prime

Minister Brian Mulroney, although he deserves the least blame because the issue was just taking shape internationally during his years in office. The leadership failures intensified greatly under Prime Ministers Jean Chrétien and Paul Martin. In Prime Minister Stephen Harper, Canadians are watching a politician move from denial, skepticism, and criticism to hurried action in less than a year. His actions so far are discouragingly inspired by too many that failed in the past, as if in panic his government had dusted off Liberal policies, given them new names and different budgets, and presented them as something fresh from "Canada's New Government." Like his predecessors, Harper has ruled out a carbon tax. His emissions targets for heavy industry have too many loopholes. And his measures do little, if anything, for the half of total emissions that do not come from large final emitters. Politically, he has apparently taken the decision to leave consumers undisturbed by serious policies, since consumers are voters. Whatever the merits of this approach politically, it tears a huge hole in the effectiveness of any serious GHG emissions reduction policy. So the fourth lesson is to develop a political smell test of the kind we will explain in a moment, so that citizens can more easily discern who is proposing serious policies and who is peddling snake oil.

Of course our politicians let us down, but perhaps we let ourselves down too. Making politicians into scapegoats lets all of us off too lightly, because we allowed ourselves to be beguiled, or we simply did not pay attention at all. Key environmental groups often did not help, peddling comforting myths that energy efficiency and renewable power – to say nothing of Suzukian prescriptions for Arcadian lifestyles – could produce huge emissions reductions, especially if these measures were reinforced by information programs and subsidies. Too many industry leaders allowed self-interest to masquerade as the public interest, spending money to support and advertise the views of a few climate change skeptics. Other leaders insisted on absolute certainty before acting, even though they would never demand that level of certainty in their own lives or in their businesses before insuring themselves against risk. Still others kept insisting that voluntary policies could reduce GHG emissions, as a way of warding off stricter pricing and regulatory policies. The fifth lesson is to beware of interest groups whose "solutions" are deficient, unrealistic, or a camouflage for self-interest.

We have often failed, again perhaps for understandable reasons, to distinguish between *policies* and *actions*. We like to talk about actions: desirable technologies such as, say, solar power, or using less energy through more efficient devices such as compact fluorescent light bulbs. Then we skip ahead in our minds to thinking these actions will be so widely taken up that GHG emissions will dramatically fall. Or we convince ourselves that with a bit more information, people will purchase these climate-saving products as a moral obligation. These changes, however, do not just happen. People take into account many factors when making decisions about technologies and lifestyles, not all of them relevant to energy use and emissions when the price of emissions is nil. Something that is free automatically gets discounted – that is, taken for granted. It becomes a non-factor for an awful lot of people.

Voluntarism can be ennobling. Voluntarism can help if consumers are willing to pay more for electricity from non-GHG-emitting generation technologies, just as it can help if people voluntarily drive slowly through school zones. But not all people drive slowly through school zones simply because it is the right thing to do; we have policies to ensure that everyone does it. We have speed limits, and fines if the limits are not respected. We back up basic, sound instincts with government policy. Why would we therefore expect to slow down or stop GHG dumping into the atmosphere without government policies to make this happen? So the sixth lesson from our past is that government policy is required to spur effective actions. The *actions* will not happen in sufficient number to make a difference unless *policies* reflect the real cost of what causes GHG emissions.

Effective policies that drive actions are also needed to avoid the plague of targetitis. Politicians, as we have seen, love to set targets. Environmentalists love targets, too. Targets make for brilliant five-second television clips and easy points of reference for the media. These days, politicians compete to set the most ambitious targets – what one commentator called "aspirational pole-vaulting."

Targets are useful, even necessary. Without them, you don't know where to aim or how fast to go. Targets are needed to measure progress or lack thereof. But targets are meaningless – they might even be counterproductive – if they're not attached to policies likely to achieve them. Years have turned into decades while targets have been missed, because policies that might have moved us toward the targets were never adopted. So the seventh lesson is that targets must be rooted in effective policies. If we can remember that targets without policies are useless, and that actions without policies will not happen in sufficient volumes to count, and if we understand which policies are effective, then, over time, we will see enough actions to meet sizable GHG reduction targets.

Today we are witnessing a welcome intensification of concern in Canada about climate change, and some reflection on past mistakes. Public awareness has certainly soared, with an Ipsos Reid poll of March 2007 showing climate change eclipsing even health care as an issue of concern. The media has caught wind of the story. Some businesses are trying to get in front of the green wave;

others are still trying to understand what has hit them, or will. Politicians are vying to outpromise one another on measures against climate change. At the ground level, however, only tentative signs at best are emerging of consumers changing buying patterns, buildings being fundamentally rethought, housing codes being overhauled, companies doing business in fundamentally different ways. The evidence of change remains fragmentary, and it will remain so until serious regulatory and pricing policies are put in place, gradually at first but more vigorously later, to bring about the necessary transformations.

We have been handicapped in Canada by a strange reluctance – with a few exceptions, and only lately – of business, union, and political leaders to take up the climate change challenge by using their public positions to alert, warn, educate, and elucidate options. No Tony Blair, the British prime minister who spoke so often and passionately about the urgency of climate change, has emerged in Canada. Nor an Al Gore. Indeed, many Canadians who watched Gore's film, *An Inconvenient Truth*, learned more than they had from any Canadian politician. Jean Chrétien devoted remarkably little political capital to talking frankly about climate change to Canadians, even though he made a hugely ambitious international commitment on Canada's behalf. Paul Martin did agree the issue was urgent, but he said that about almost every issue, a habit which, over time, caused listeners to believe that nothing was actually that important. Stephen Harper, as is his wont, has been more vigorous in underscoring the failure of his Liberal rivals than in explaining the perils and imperatives of global warming. His speeches on this, as on other matters, have tended to be matter-of-fact, utilitarian, descriptive, and dull.

It is useful, however, for Canadian governments to try to educate Canadians about climate change, at the very least about its long-term impact on their country and what they can do about it in their own lives. Actions do count: acquiring a low- or zero-emissions car, driving less, purchasing more energy-efficient appliances, insulating homes, paying extra where possible for zero-emissions electricity from renewables, using biofuels, pressing politicians and officials to take climate-friendly decisions in everything from local urban and building design to national policies. These actions are important, in and of themselves, and for the general awareness they help raise – awareness that establishes more support for the effective policies that, over time, will produce many more of these kinds of actions. To repeat: voluntarism and subsidies will not bring about the sharp reductions in GHG emissions required. They can help at the margin, set examples for others, and create the right public climate for implementing the economic policies that will really drive actions to curb emissions.

What should we be on guard against as Canada tries to make up for wasted years and get serious about reducing GHG emissions? How can we distinguish between what Thomas L. Freedman has shrewdly called "energy politics" and "energy policy"? What smell tests should we use for distinguishing serious talk from false promises among politicians (and environmentalists and business associations)? Here is a checklist.

1. If targets are proposed, but the politicians setting them do not detail how they will be reached, assume failure.
2. If, after advancing a target, politicians propose policies that lack a regulatory or financial constraint on GHG emissions, assume failure.
3. If politicians insist that behavioural change by individuals alone will solve the climate change problem, assume failure.
4. If politicians talk only about exciting, enticing actions – renewable energy, energy efficiency, carbon capture and

storage – but say nothing about economic policies to force these actions, assume failure.

5. If politicians complain about jurisdictional constraints, assume failure.
6. If politicians crisscross the country or their province handing out subsidies and offering photo opportunities of themselves in front of wind turbines, research laboratories, or corn fields, assume failure.
7. If politicians make a big deal for ideological or other reasons about ruling out all forms of GHG taxes, and don't substitute for those tax policies ones that rely on market-oriented regulations such as emissions caps with tradable permits, assume failure.
8. If politicians insist that Canada can meet its Kyoto commitment, offer the benefit of the doubt that they are not lying, just being disingenuous.

We need to add some general comments to this list of political smell tests. Effective GHG emissions reduction policy must apply to the entire economy. Industry cannot be the target of serious economic policies while consumers are exempt, and vice versa. The challenge of climate change is such that all parts of Canadian society need to contribute. So, too, all regions have to contribute. Although producers of fossil fuels are concentrated in a few provinces, consumers are everywhere, as are businesses and utilities. In a country such as Canada, debates will flare about which industry or region must do what, and which formula for sharing adjustments should be applied to consumers and industry. Some of this debate is normal democratic dialogue; some of it will be particularly Canadian. That there will be debate and disagreement about internal obligations should not be a deterrent to act, because they come with the territory, as it were.

Canadians need to become realistic after years of failure and fantasy. We have to choose among four policy options. Each can be made to work. We need to combine what we might call quantity restraints on the total of GHGs emitted with price signals to deter or prevent those emissions.

We can use a GHG tax. We can use an upstream carbon cap-and-tradable-permit system that restricts the carbon flow of industries by requiring them to acquire permits for the carbon they process into refined petroleum products for consumers or electricity generation plants. This approach will ripple through the economy, pushing along decarbonization, because the resulting higher prices will induce some fuel switching.

We can use a downstream cap-and-tradable-permit system, akin to the large final emitters system toyed with by Conservatives and Liberals. It can be based on emissions per unit of industrial output (intensity) or absolute quantity of emissions. The stringency of the cap counts. Intensity policy can work, but not if the intensity target becomes overwhelmed by a surge in units of output, as in the oil sands. A policy that targets only large final emitters allows half the country's emissions off the hook. For this downstream system to work, it has to be accompanied by a similar emissions cap-and-permit system for final energy consumption in vehicles, homes, offices, and small industries, or a GHG tax for these energy consumers.

Finally, we can use a carbon management standard. It works its way down through the economy because it is imposed on producers that are obligated to provide certain low- or zero-emissions products. It could also apply as a carbon portfolio standard for large emitters of GHGs. And we will need command-and-control policies and targeted market-oriented regulations for such areas as vehicle emissions requirements, appliance regulations, and building codes.

Canadians wanting serious action can choose from the four main policy options, or blend some of them. The specific combination is less important than getting Canadians to understand that this basket of policies – *and only this basket of policies* – will produce significant GHG reductions in future decades. Modalities are important, even critical, but nothing is more essential at this point than understanding which are the serious policies and which ones are certain to keep failing. Each of these serious policies can be made to work, and the sooner we put them in place the better. We have lost so much time; we can ill afford to dither any more. And remember that even if we begin implementing serious policies tomorrow, their impact will not be felt for another decade or so. The longer we delay, the later we will see results, because there are no quick fixes to our climate change challenge.

We need to implement serious policies quickly but tighten them gradually. It would be stupid to trigger an economic recession, or grievously hurt particular industrial sectors, because we are rushing to meet a short-term Kyoto target. Most scientists are telling us that although the hour is late, if countries can make substantial changes so that emissions are in rapid decline in a few decades, then we will have acted in time to slow down, and eventually reverse, GHG buildup in the atmosphere. Technological changes and breakthroughs take time to invent, refine, and deploy in the market. When they hit the market and are disseminated, they make an impact quite rapidly and can then be transferred or sold to developing countries.

We cannot predict what technological changes will reshape the future. We have shown that this basket of policies will drive us toward low- and zero-emissions technologies, but a variety of possible actions can come out of these policies. Thinking well down the road, people could dramatically reduce their car use, although not with the ones they own today. They could drive much smaller cars. They could drive hydrogen cars, probably using a fuel cell but perhaps combusting the hydrogen. They could drive biofuel cars. They could drive plug-in hybrids that use a small amount of diesel, gasoline, or biofuel, but are mostly powered by electricity from the grid. We don't know the exact technologies and the energy sources that will feed the grid in the low- and zero-emissions future. And we don't need to know. Provided we put the right policies in place that send the unmistakable signals to businesses and consumers, with the necessary lead times for adapting, then markets will figure out the technologies and fuel sources. Markets, after all, are businesses and consumers making choices that work best, given their own preferences and budgets – and given the laws imposed by governments, including the policies Canadians should enact to protect the atmosphere for others and themselves.

Nor can we predict how much energy Canadians will use in the future. Certainly those who favour energy efficiency as the prime answer to our emissions challenge believe that therein lies the key to decarbonization. The likelihood of energy efficiency holding the key, however, is quite low. New energy-efficient equipment is often more expensive, and consumers will not necessarily and voluntarily lay out more money for it. We live in a free-market, consumer-driven society that has brought us many advantages and is unlikely to change fundamentally. Underlying market economics are impulses that drive humans everywhere – for mobility, comfort, food, status, leisure, entertainment, and gift-giving. Yes, it is possible that we could consume less, but this, too, is highly unlikely in a modern, industrialized economy – to say nothing of the developing world where the bulk of the population lives and where people would like to enjoy some of the benefits we do. In other words, the prospect of Canadians returning to a simpler, Arcadian, smaller-footprint life is as unlikely as people in the developing world being content for their children to remain in the conditions in which they find themselves. Even if we in Canada could imagine that we would rediscover the joys of the simple life, it is illusory and, frankly, insulting to imagine that those in the developing world would accept less than something better for their societies.

There are dangers to the environment from Canada's consumer society as it has evolved and with the laws it has created, which is why new ones are needed to curb GHGs. But it is dreaming in Technicolor to believe Canadians will use 10 per cent less energy in each of the next four, five, or six decades, and it is a certainty that people in the developing world will use more than they do today. So if energy use will rise – in Canada and beyond – then technologies to curb emissions, plus different fuels, are an urgent necessity.

There will be dislocations for individuals and businesses along the way as they adapt to changed policies. Long lead times can ease the transition. In general, as we have seen, the Canadian economy will do fine. We have considerable renewable energy potential from hydro power, wind, solar, biomass, geothermal, and ocean sources. We have immense technical capabilities, since the demands of geography and climate have forced us to become proficient in resource technology and long-distance transportation. We need to implement the right policies to turn those capabilities in new directions to exploit our vast fossil fuel reserves and natural gas differently.

Policies can be implemented that do not harm the economy of our fossil-fuel-rich regions such as Alberta. A low- or zero-emissions economy does not have to be biased against any particular form of energy. Money does not have to be extracted from Albertans and redistributed around the country, if GHG tax revenues are returned proportionately to the regions in which they are raised. Oil sands development might occur slightly more slowly than in recent years, but given the recent madcap pace of development with its attendant pressures on local municipalities, school boards, and labour supply, slightly slower would mean better. And Alberta, with the right policies, visionary political leadership, and business acumen, could marry its ability to develop resources with the new imperative of reducing emissions and so become the world leader in sustainable development.

Other countries are already ahead of us. The United Kingdom, one of our closest allies, has a GHG tax, an emissions cap, tighter energy regulations, and support for alternative energies. It has a national mission, endorsed by both major political parties, to lower emissions by 60 per cent by 2050. That is the country's target and, having heeded warnings about targets without policies, the United Kingdom has policies to get most of the way there and is working on the rest. Norway and other Scandinavian states are moving along the same path. The European Union has given itself ambitious targets and forceful policies. Parts of the United States are adopting policies that will lead to climate-friendly actions. We can reasonably expect in the next few years significant national actions in that country.

So Canada, in adopting policies proposed here, will not be on a fool's errand, wandering around alone in search of an unattainable goal, sniggered at by others, crippling itself in the process. Future international agreements and intensified efforts are coming. This time, having learned from its sorry past, Canada needs to be part of those agreements, not just rhetorically but substantively in the targets we accept and the serious policies we adopt to meet them. We can be among the leaders, so that the deeds of a self-imagined moral superpower will finally speak louder than words. We need to act now. The time for hot air is over.

Update to Paperback Edition

If, then, Canada needs to act now and the time for hot air is over, can we be optimistic that our governments and industries – and the Canadian public – have grasped the urgency of moving to reduce greenhouse gas emissions? Have we understood that only compulsory policies can be effective, or are we still wedded to the politically easy but ineffective policies of the past? Rhetorically, the business community has come a long way from its years of denial and delay, but can we be sure that the new expressions of concern and calls for action are not a public relations camouflage?

The actions taken recently in Canada, and those that can be reasonably anticipated based on governments' statements of intent, plus developments in the United States and beyond, suggest that policy is moving like a crab – sideways, with occasional retreats, bursts of forward motion, and a general zigzag direction towards the goal of lower absolute emissions.

Governments are proposing to implement, and the business community is accepting, cap-and-trade markets for carbon, carbon management standards, and pricing for carbon. The atmosphere is finally not just something into which we can freely dump carbon. Even what had been the political heresy of heresies – a carbon tax – has been implemented in Quebec and British Columbia. It has also been endorsed by prominent economists, and not ruled out by leading business interests. For many politicians and citizens, a carbon tax still remains a bridge too far, but the bridge is coming into view.

Good news is also arriving in the changed political discourse. No political party, business group, or civic association now rejects the seriousness of climate change. Deniers and skeptics are still among us but they have been relegated to the cheap seats of the democratic public square where people discuss issues and thrash out alternatives. The fact that no political actor tries to appeal to them anymore speaks to their marginality.

As we have seen, the Conservative Party and its federal antecedents used to throw up every imaginable objection to taking climate change seriously. Now the party is committed to reducing Canada's emissions by 20 per cent by 2020. Based on our analysis of their own proposals and those of the provinces, we believe that Canada will not reach this target; nonetheless, it would have been inconceivable just a few years ago for the Conservative Party to have dreamed of setting this goal for Canada. But it is important to remember that this 20 per cent reduction target uses 2006 as its baseline, not the 1990 baseline of previous governments. During those 16 years, emissions rose by about 27 per cent. The Conservatives argue the Liberals were to blame for this increase and insist they are not going to jeopardize the Canadian economy trying to make up for "Liberal" emissions. The atmosphere, however, is indifferent to "Liberal" or "Conservative" emissions.

Nonetheless, at the federal and provincial levels of government there is a deeper willingness to act on climate change compared with a few years ago. Joint federal–provincial action, however, remains elusive. Canada's federal system of government presents the advantage of provincial experimentation from which others can learn, but also the complications of governments getting their acts together. British Columbia and Alberta, for example, are moving in very different ways; British Columbia towards aggressive emission reduction targets and a carbon tax, Alberta has planned modest long-term reduction targets but actual emissions increases for several decades and, of course, no carbon tax. British Columbia and Alberta are at the policy extremes of the national debate. The provinces and territories collectively have failed twice to reach a common position, and the other big fossil-fuel producing provinces of Saskatchewan and Newfoundland are clustered together around Alberta's go-slow position.

The federal Conservative government, observing this interprovincial disarray and not wishing to fight with Alberta, has adopted a hands-off approach, essentially encouraging each province to do its own thing. The result, of course, is a mishmash of policies across the country, and the strong likelihood that Canada will not meet its target of a 20 per cent reduction by 2020. Should the target be missed, Canada would once again stand exposed as a purveyor of political hot air – a country that talks a better game than it plays.

According to a legal opinion prepared by one of the country's leading constitutional lawyers, Peter Hogg, the federal government has an array of powers that would allow it to supersede provincial environmental regulations. It could be – but this remains only a possibility – that if Canada's international obligations were frustrated by Alberta's refusal to take more vigorous action, a federal government could choose to use those powers. Any such attempt, of course, would be met with a furious political and legal reaction from Alberta and, presumably, other fossil-fuel-producing provinces. Such are the dilemmas of Canada being a federation and a fossil-fuel producer.

We showed earlier in the book why the first policies announced by the Conservatives, after the demise of the hapless Rona Ambrose and the arrival of John Baird as environment minister, would slow the growth of emissions but not reduce them by 20 per cent from 2006 levels by 2020. We estimated that emissions in 2020 would remain at about 2006 levels – better, to be sure, than business as usual – but not nearly good enough.

The government had exaggerated the reductions because of a loophole that allowed large industrial emitters to subsidize apparent reductions in the uncapped and untaxed emissions of all the other sectors of the economy. The flaws in the program, and the gap between the government's target and its unlikely realization, were noted by the Conference Board, the TD Bank, the Pembina Institute, and the National Round Table on the Environment and the Economy.

Stung, perhaps, by the criticisms, the government refined its proposals. The government remained wedded to its intensity reduction model in the early years – 18 per cent over three years – followed thereafter by a yearly drop of 2 per cent. Companies could still pay into a technology fund at $15 per tonne for missing the target, with that option being phased out by 2017. The government did get more serious, however, in setting intensity improvement requirements tied to individual oil sands processing and coal-fired generation plants. As noted earlier in the book, the oil sands pose a serious challenge for reducing GHGs in Canada. Demand for the oil locked in them is enormous; the capital outlays to exploit them are prodigious; the revenues that flow to governments are irresistible; and the jobs created are lucrative. Statistics Canada reported the tectonic shift that capital investment in the oil sands in 2008 exceeded capital investment for the country's entire manufacturing sector. As a result, the oil sands have become the biggest single economic investment driver in the Canadian economy, with the prospect of oil above $100 a barrel providing healthy profits for companies, despite higher costs for labor, supplies, and royalties than most of them had anticipated when they bid for exploration rights. Still, the oil sands are bad carbon polluters, much worse than conventional oil.

The federal government, and Alberta, are making a huge bet on technology and are placing most of the responsibility for reducing overall emissions on industry rather than anticipating changes in consumer behavior. This is, of course, the path of the least political resistance. They are assuming that facility-specific emissions intensity requirements can be met by new or improved low-emission technologies over the next decade. After that, they are placing their hopes in carbon sequestration which on its own will be unlikely to absorb all the anticipated reductions. Then, there are the offsets that companies can buy; in other words, they can spend money in other sectors of the economy while doing little at their own facilities to reduce emissions. Any government would achieve more success if it put a tight cap on allowances for offsets; or, better still, disallowed them altogether. Rather than restrict a cap-and-trade system to some sectors of the economy, it would be better to extend it to the entire economy, or use a carbon tax to cover all remaining unregulated emissions.

A uniform economy-wide price for GHG emissions helps society by not forcing high-cost emissions reduction in some sectors while ignoring low-cost actions in others. This approach of the federal government turns out to be a more costly way economically of reducing emissions – but not necessarily less costly politically, which is why the federal government, Alberta, and other risk-adverse governments follow it. Emissions for backyard patio heaters, and thousands of other energy-consuming and carbon emitting devices, are free under the Conservatives' approach. The same goes for thousands of small industrial plants. With these and other exemptions and provisions, federal policy is more costly to the economy than need be.

There is also a philosophical contradiction at the heart of the Conservatives' approach. Their preference for using taxes to achieve social objectives invariably trumps other tools such as direct spending, new government programs, subsidies, and regulations. Railing against "red tape," "bureaucracy," and "unnecessary regulations" has been a staple in conservative political discourse, and for the Harper Conservatives.

And yet, the Conservatives have ruled out a carbon tax in favor of extremely complicated regulations for the offset system, technology-specific provisions, and industrial and facility targets. Their approach will necessitate exactly what they usually rail against: lots of bureaucracy, rules galore, and regulations only lawyers and accountants will love. Administrative costs are likely to be substantial. Even a carbon cap-and-trade system generates an army of brokers and traders for permit exchanges. These kinds of costs are not required by a simple carbon tax – universally applied, transparent, and clearly understood.

Politics obstructs the Conservatives' view. They have denounced any mention of a carbon tax as a "Liberal" plan, a "tax-and-spend" scheme that will slam ordinary Canadians. With their party pledged to reduce taxes, the mere mention of a new tax sends shivers up Conservatives' spines, even if the tax revenue were offset by lower taxes elsewhere. As a result, Canadian national climate change policy remains where Prime Minister Jean Chrétien left it in 1994 when he went to Calgary and told everyone there would be no carbon tax while he was prime minister. The Chrétien pledge remains a cornerstone of Conservative policy.

The federal Conservatives' cousins in Alberta, a province without a sales tax and the lowest personal income taxes in

Canada, share this antipathy to a carbon tax. They also retain a prickly determination to control all aspects of natural resources, including pollution such as GHG emissions that respects neither provincial nor national borders. They remain suspicious of the intentions of other Canadians who may somehow curtail Alberta's control of its natural resources, and grab the rents from them, possibly by trading Alberta emissions outside the province. And the Alberta Conservatives are determined that nothing, not even a slight pause for reflection, will slow down the province's hell-bent-for-development approach to the oil sands.

The result, predictably, is a set of "made in Alberta" policies that will allow emissions to rise for many years and, over time, will damage the province's reputation and be overwhelmed by developments outside the province. Defenders of Alberta's modest approach applaud its honesty, insisting Alberta will meet these targets, unlike other provinces and Ottawa that establish impressive targets but fail to deliver. Defenders insist the technological fixes necessary to restrict emissions sooner are just not available. Critics note that the Alberta approach certainly is modest – so modest that, if unchanged, it will cause the province to be seen as an emissions pariah within Canada, North America, and internationally.

In 2007, Alberta introduced the country's first cap-and-trade regulatory system on major emitters. It was intensity based. Each industrial emitter was supposed to either lower its emissions per dollar of production by 12 per cent from 2006 levels by July 2008, buy credits from another company in Alberta that had exceeded this target, or purchase offset credits – but only from elsewhere in Alberta. The government wanted to keep money in the province. It did not want money for offsets or trades going to Toronto or Montreal, a position rooted in Alberta's traditional wariness of Central Canada, but one almost certainly destined to be overtaken by forthcoming international developments.

Emitters are only paying for what they emit above the Alberta intensity target. Remember that companies had been improving their energy intensity anyway because it made good business sense to use less energy per unit of production. The $15-per-tonne charge for emissions over the limit will in practice work out to about $3 per tonne when averaged over all emissions, and under some circumstances will be even lower. The Alberta policy will therefore have only a minimal effect on companies, especially in the fast-growing oil sands industry.

The trouble for the environment – if not for companies' bottom lines and the Alberta government – is that intensity reduction targets will be overwhelmed in the oil sands sector when overall output skyrockets. Intensity reduction targets only slow down the rate of increase of emissions; they do not reduce those emissions in absolute terms if overall production is soaring. Even if, as the Alberta government proposes, subsidies are offered for carbon capture and storage and for renewable energy, and even if municipalities are given assistance to plan low-emission urban development, and tighter energy efficiency standards are imposed, Alberta's emissions will increase by about 20 per cent by 2020. This is the government's own projection. The increase will be lower than had nothing been done, but it remains a very large increase. Moreover, the Alberta government's target for 2050 calls for only a 14 per cent reduction in absolute emissions – a target far below those set by other industrialized places, and ominously far from the Harper government's nominal target of a 50 per cent reduction by 2050 for Canada.

Canada cannot meet that 50 per cent target if Alberta, the country's largest per capita emitter, manages only a 14 per cent reduction. The arithmetic does not add up. It is true that Canada's target cannot be met if other provinces, notably Ontario, do not do more than they have thus far announced, but no province is as responsible for so many per capita emissions as Alberta. Whether Canada meets its target depends on all provinces, but more on Alberta than any other. Our modeling, for example, shows that for Canada to reach a 50 per cent reduction by 2050, Alberta's emissions must fall by 40 per cent – not 14 per cent.

Alberta's approach leads to the tragedy of missed opportunities. The province could lead the world in demonstrating how to marry development of fossil fuels with lower emissions; in other words, sustainable development. It could stage the development of the oil sands to allow technology to catch up and to prevent local economies for labor and materials to overheat. It could have a much more aggressive policy for the construction of a carbon pipeline and for sequestration projects. It could weave a carbon tax into the provincial economy, and design it to keep the revenues within the province if local politics demanded such a policy. Instead of thinking big and thinking ahead, Alberta has settled for a head-in-the-sand approach that will leave Alberta behind in the race for new technologies, stigmatized in the advanced industrialized world as the place with the worst emissions record. What remains lacking in Alberta is a political (and business) vision that does not tap into the province's fears but into its ambitions.

By contrast, the British Columbia government under Premier Gordon Campbell became the North American leader in designing effective climate change policies. Years from now, his government's climate change policies will be the template for other governments wanting to take effective action. What the British Columbia government is doing to reduce GHG emissions, with all the attendant political risks, will be viewed in climate change policy with the same respect as Premier T.C. Douglas's introduction of publicly-financed health care in Saskatchewan. With the 2008 British Columbia budget that introduced a carbon tax and a series of other measures, the policy ground shifted in Canada.

The taboo of a carbon tax vanished. Political fears surrounding serious action against emissions weakened, although governments across Canada watched to see whether Premier Campbell would be strengthened or wounded by his dramatic proposals.

Campbell's first term, 2001 to 2005, featured little interest in climate change. By cutting spending and taxes in the first term, Campbell developed a reputation as something of a right-winger, despite leading the provincial Liberal Party. It seemed improbable that he should become interested in climate change, an issue previously talked about mostly by leftish politicians and commentators. He not only became interested, he got religion, as they say, by reading, consulting, and traveling and decided that global warming posed a serious long-term threat to humanity, and to his province. After all, he needed only to fly over the interior of British Columbia to observe the devastation caused by the mountain pine beetle that had flourished in part because no cold winters had blunted their advance.

So Campbell set the boldest targets in Canada: 33 per cent emissions reductions by 2020 from 2007 levels, and at least 80 per cent by 2050. He did not know how or if his province could meet those targets, so in this sense he was like other Canadian politicians who had set targets without proposing serious policies to achieve them. Campbell, however, established a climate change secretariat attached to his office with a deputy minister in charge, signaling that climate change would be a central priority.

While the secretariat prepared plans in 2007, the government made its first dramatic announcement: that B.C. Hydro must ensure that at least 90 per cent of new electricity generation comes from clean sources. The policy stipulated that coal plants would not be allowed unless they included carbon capture and storage technology that cut emissions almost to zero. With this electricity policy alone, the Campbell government distinguished itself as a leader in Canada, North America, and even globally.

The provincial budget of February 2008 – the Medicare budget for climate change policy – introduced a carbon tax as the centerpiece of a suite of policies. Quebec had earlier earned the distinction of being the first Canadian province to levy a carbon tax, but their tax was not designed to produce changes that would reduce emissions. The levy was tiny – $3 per tonne on producers. The Quebec government said it did not expect the tax to work its way through to consumers. It was not designed to change technology choices, but rather to send a signal; it was more politics, in other words, than substantive policy. It was a cash grab, too.

By contrast, the British Columbia carbon tax was set at $10 per tonne in 2008. Each year until 2012, the tax will rise by $5 per tonne until it reaches $30 per tonne. The tax will produce a 2.4-cent-a-litre rise in the price of gasoline in 2008, rising to 7.5-cents-a-litre by 2012. The tax is expected to raise $1.85 billion in the first three years. Every penny, by legislation, must be returned to taxpayers in the form of lower personal and corporate taxes and tax credits for low-income people.

Three essential principles of effective carbon tax policy are evident. First, the carbon tax begins at a low rate but rises steadily, so that people are forewarned and have time to adjust to new circumstances. Second, all the money raised by the new tax is recycled into tax reductions elsewhere. The net effect on government revenues is neutral; that is, the carbon tax cannot be considered a tax grab, and will have marginal macroeconomic impact. Third, the recycled tax money is weighted to personal income tax reductions and credits, especially for low-income people, to make the tax more politically palatable. Indeed, the government even sent an initial $100 cheque to every British Columbian to "help begin the transition to a lower carbon lifestyle."

British Columbia's GHG emissions in 2005 were 66 million tonnes. Reducing this total by 33 per cent, as the Campbell government pledged, will mean a 22-million-tonne reduction. And yet the government estimates that this carbon tax, with its schedule of $5-per-tonne yearly increase to $30 per tonne, will reduce emissions by only 3 million tonnes, although this is a conservative estimate.

The conclusion is two-fold. First, the carbon tax must and will, politics permiting, go much higher. Second, a range of other policies is needed to supplement the carbon tax, which is exactly what the Campbell government proposed. It is investigating a cap-and-trade system for big emitters. It is spending $14 billion on public transit in the Lower Mainland. It is going to introduce vehicle emissions standards. It is introducing a Green Building Code that will be the most rigorous in Canada. In short, through these and other measures, British Columbia is using the only policies that reduce emissions: compulsory policies introduced irrevocably but increased gradually to produce the incentives and disincentives to move the economy towards a less carbon-emitting future. British Columbia, in other words, wants to lead not just within Canada, but in North America; Alberta is content to follow at the rear.

Other provinces' commitment to GHG reductions remains rhetorically strong, but their deeds fall somewhat short of their words. In Ottawa, the government has no plan to attempt a national policy, for it prefers to preach the virtues of "open federalism," which means no search for national norms or solutions, except equalization. But even after introducing changes to that program, it brokered a side deal with angry Nova Scotia.

Saskatchewan, happily, is requiring that new coal-fired plants sequester carbon, but the province is leery about jeopardizing its own booming fossil-fuel industries by adopting other serious measures. Ontario frets about its car industry, failing to realize that Japanese producers, having already established themselves as brand leaders for lower-emissions vehicles, will be clobbering North American competitors when tougher vehicle emissions standards arrive, as they will. Again, rather than adopting policies that will get them ahead of the green curve, Ontario (like Alberta) is playing a defensive game, hoping to slow things down.

Slowing down the movement towards climate change action in Canada is short-sighted for environmental reasons but also for political ones. Canada has a fundamental climate change choice in North America even if Canadians do not yet recognize it. We can try to influence continental policies from their inception, or we can wait on events and then, in typical Canadian fashion, agonize over whether to negotiate arrangements on terms already set by the United States.

About half the U.S. states are already moving against climate change, as are hundreds of municipalities. Some of these jurisdictions have moved beyond most policies contemplated or introduced in Canada, except for in British Columbia. This is one of the reasons why it comes as a shock to Canadians' overweening sense of moral superiority to learn that post-Kyoto, GHG emissions actually rose slightly more slowly in the U.S. than in Canada. Congressional action will certainly come in 2010–2012, because the Democratic Party is committed to aggressive policies, as is part of the Republican Party led by John McCain. It is overwhelmingly likely that, after all the to-ing and fro-ing that is inherent in the U.S. political process, and despite all the lobbying from the dwindling band of legislators and interests seeking as little action as possible, the U.S. will emerge with a national cap-and-trade system to lower carbon emissions, and vehicle emissions standards more or less modeled on the ones already crafted in California. It makes eminent sense for the U.S. to create national markets and standards to replace the patchwork of state provisions that impose costs on companies.

Once the Americans move, their actions will mock Alberta's decision to restrict trading in carbon within the province. Every Alberta company – especially all the fossil fuel producers – will be clamoring for trading access to the bigger market. Worse, if Alberta does not change its policies, oil from the tar sands might be targeted by jurisdictions that will refuse to allow the product to be sold there because it is so polluting. For example, if the U.S. insists China agree over time to limits on its GHG emissions, or else the U.S. will employ trade sanctions against it, might the same sort of approach not be used against GHG emissions from the tar sands?

A much more forward-looking approach in Canada would be to get ahead of the forthcoming U.S. measures: to establish best-practice standards for North America, test them in Canada, and be in on the ground talking to the Americans as they establish their new policies, working to develop the most aggressive yet workable arrangements based, in part, on what their most active states and Canada will have already learned.

The ostrich approach, favoured by a strange alliance of Alberta laggards and anti-American Canadian nationalists, would be to believe that Canada is already doing all it can, and will not be "dictated to" by the United States. It might even insist (against the evidence) that our standards are better, and that we should, therefore, go it alone in North America. Because Canada cannot get its own act together with national standards or markets or taxes to combat climate change, we risk moving unprepared into a North American policy environment shaped entirely by Americans. And, of course, we handicap ourselves in international negotiations if other countries know that the commitments of the national government will be so much hot air when confounded by provincial recalcitrance.

As we have argued, this is the time for bold action, vision, and commitment, some of which is now happily apparent. But much more is needed to bring about the de-carbonization of Canada. Since the last report of the IPCC, even more scientific evidence has accumulated about global warming, much of it suggesting that the IPCC might have underestimated the pace and scope of warming. So the gap is widening, alas, between the urgency that science underscores, and the slowness of the political response.

There will be mistakes of analysis, policy, and execution in the years ahead. There will be trials and errors. It is always thus when societies enter on a new era. We are entering, at different paces and with varying enthusiasms, a new era of production and consumption in order to move away from behaviour that warms the atmosphere. The Industrial Revolution that began the long period of significantly increased human contributions to global warming stretched over decades, producing products, changes in lifestyle, political reconfigurations, economic theories, and practices that could not have been imagined when the Revolution started, for example, steam produced from burning coal.

So it will be with de-carbonization. The struggle towards that goal will stretch over decades, sprawl over many aspects of human activity, produce many setbacks and disappointments, but also bring about changes in behavior, technological advances and product discoveries that no one can imagine today. No one can outline precisely every step in this long path, but we need to get on that path, and stay there, using policies that we believe can work, adjusting them as required, and insisting that those whom we elect attend to this global challenge in the geographic space called Canada with the dedication it deserves.

References

Chapter 1

Arctic Council, 2004, "Arctic climate impact assessment," New York: Cambridge University Press.

Canada, 2006, "The mountain pine beetle: a synthesis of biology, management, and impacts on lodgepole pine," Canadian Forest Service, Natural Resources Canada.

Canadian Council of Ministers of the Environment, 2003, "Climate, nature, people: indicator's of Canada's changing climate," Winnipeg: Canadian Council of Ministers of the Environment.

Huebert, R., 2001, "Climate change and Canadian sovereignty in the Northwest Passage," *ISUMA*, 2, 4.

Intergovernmental Panel on Climate Change, 2007, "Fourth Assessment Report, Climate Change – The Scientific Basis," Paris: World Meteorological Organization, United Nations Environment Program.

Kaiser, J., 2002, "Breaking up is far too easy," *Science*, 297, 1494-1496.

National Academy of Science, 2002, "Abrupt Climate Change: Inevitable Surprises," Committee on Abrupt Climate Change, National Research Council, US.

Nikiforuk, A., 2007, "Pine Plague," *Canadian Geographic*, January/February.

Stern, N., S. Peters, V. Bakhshi, A. Bowen, C. Cameron, S. Catovsky, D. Crane, S. Cruickshank, S. Dietz, N. Edmonson, S.-L. Garbett, L. Hamid, G. Hoffman, D. Ingram, B. Jones, N. Patmore, H. Radcliffe, R. Sathiyarajah, M. Stock, C. Taylor, T. Vernon, H. Wanjie, and D. Zenghelis, 2006, "Stern Review: The Economics of Climate Change," London: H.M. Treasury.

Chapter 2

Bernstein, S., 2002, "International institutions and the framing of domestic policies: The Kyoto Protocol and Canada's response to climate change," *Policy Sciences*, 35, 203-236.

Canada, 1990, "Canada's Green Plan for a Healthy Environment," Ottawa: Government of Canada.

Canada, 2000, "Government of Canada Action Plan 2000 on Climate Change," Ottawa: Government of Canada.

Harrison, K., 2006, "The struggle of ideas and self-interest: Canada's ratification and implementation of the Kyoto Protocol," Annual Meeting of the International Studies Association, San Diego, California, March 22-26, 2006.

Smith, H., 1998, "The provinces and Canadian climate change policy," *Policy Options*, May, 28-30.

Chapter 3

Canada, 2002, "Climate Change Plan for Canada," Ottawa: Government of Canada.

Gelinas, J., 2006, "Report of the Commissioner of the Environment and Sustainable Development to the House of Commons", Ottawa: Office of the Auditor General of Canada.

Goldenberg, E., 2006, "The way it works: inside Ottawa," Toronto: Douglas Gibson Books, McClelland & Stewart.

Chapter 4

Canada, 2005, "Project Green: Moving Forward on Climate Change: A Plan for Honouring our Kyoto Commitment," Ottawa: Government of Canada.

Intergovernmental Panel on Climate Change, 2005, "IPCC special report on carbon dioxide capture and storage," New York: Cambridge University Press.

Jaccard, M., N. Rivers, and M. Horne, 2004, "The Morning After: Optimal Policies for Canada's Kyoto Commitments and Beyond," CD Howe Institute Commentary, No. 197.

Jaccard, M., J. Nyboer, and B. Sadownik, 2002, "The Cost of Climate Policy," Vancouver: UBC Press.

Chapter 5

Lovins, A., 2005, "More profit with less carbon," *Scientific American*, September.

Torrie, R., R. Parfett, and P.Steenhof, 2002, "Kyoto and Beyond: The Low Emissions Path to Innovation and Efficiency," Ottawa: The David Suzuki Foundation and The Canadian Climate Action Network.

Chapter 6

Bataille, C., M. Jaccard, J. Nyboer, and N. Rivers, 2006, "Towards general equilibrium in a technologically rich model with empirically estimated behavioural parameters," *The Energy Journal.* Special issue on hybrid modelling of energy-environment policies.

Canada, 2005, "Project Green: Moving Forward on Climate Change: A Plan for Honouring our Kyoto Commitment," Ottawa: Government of Canada.

Canada, 2007, "Regulatory Framework for Air Emissions," Ottawa: Government of Canada.

Dion, S., 2006, "Building a sustainable future for Canada: Stéphane Dion's energy and climate change plan," Stéphane Dion Campaign.

Jaccard, M., N. Rivers, C. Bataille, R. Murphy, J. Nyboer, and B. Sadownik, 2006, "Burning our money to warm the planet: Canada's ineffective efforts to reduce greenhouse gas emissions," CD Howe Institute Commentary, 234.

Chapter 7

Stavins, R., 2001, "Experience with market-based environmental policy instruments," in "The Handbook of Environmental Economics," eds., Maler, K.-G. and J. Vincent, Amsterdam: North Holland/Elsevier Science.

Nordhaus, R., and K. Danish, 2003, "Designing a mandatory greenhouse gas reduction program for the US," Washington: Pew Center on Global Climate Change.

Chapter 8

None

Chapter 9

None

Chapter 10

None

Acknowledgements

From Jeffrey: I would like to thank the editors of the *Globe and Mail* for continuing to allow me the privilege of writing my national affairs columns, many of which have dealt with climate change.

From Mark and Nic: Our thanks go to Rose Murphy, John Nyboer, Chris Bataille, and Jotham Peters.

From all: Our thanks to Barbara Czarnecki, our splendid copy-editor.